Communications in Computer and Information Science 2243

Rationale

The CCIS series is devoted to the publication of proceedings of computer science conferences. Its aim is to efficiently disseminate original research results in informatics in printed and electronic form. While the focus is on publication of peer-reviewed full papers presenting mature work, inclusion of reviewed short papers reporting on work in progress is welcome, too. Besides globally relevant meetings with internationally representative program committees guaranteeing a strict peer-reviewing and paper selection process, conferences run by societies or of high regional or national relevance are also considered for publication.

Topics

The topical scope of CCIS spans the entire spectrum of informatics ranging from foundational topics in the theory of computing to information and communications science and technology and a broad variety of interdisciplinary application fields.

Information for Volume Editors and Authors

Publication in CCIS is free of charge. No royalties are paid, however, we offer registered conference participants temporary free access to the online version of the conference proceedings on SpringerLink (http://link.springer.com) by means of an http referrer from the conference website and/or a number of complimentary printed copies, as specified in the official acceptance email of the event.

CCIS proceedings can be published in time for distribution at conferences or as post-proceedings, and delivered in the form of printed books and/or electronically as USBs and/or e-content licenses for accessing proceedings at SpringerLink. Furthermore, CCIS proceedings are included in the CCIS electronic book series hosted in the SpringerLink digital library at http://link.springer.com/bookseries/7899. Conferences publishing in CCIS are allowed to use Online Conference Service (OCS) for managing the whole proceedings lifecycle (from submission and reviewing to preparing for publication) free of charge.

Publication process

The language of publication is exclusively English. Authors publishing in CCIS have to sign the Springer CCIS copyright transfer form, however, they are free to use their material published in CCIS for substantially changed, more elaborate subsequent publications elsewhere. For the preparation of the camera-ready papers/files, authors have to strictly adhere to the Springer CCIS Authors' Instructions and are strongly encouraged to use the CCIS LaTeX style files or templates.

Abstracting/Indexing

CCIS is abstracted/indexed in DBLP, Google Scholar, EI-Compendex, Mathematical Reviews, SCImago, Scopus. CCIS volumes are also submitted for the inclusion in ISI Proceedings.

How to start

To start the evaluation of your proposal for inclusion in the CCIS series, please send an e-mail to ccis@springer.com.

Sanju Tiwari · Fernando Ortiz-Rodriguez ·
Miguel-Angel Sicilia · Tek Raj Chhetri
Editors

Artificial Intelligence: Towards Sustainable Intelligence

Second International Conference, AI4S 2024
Alcala de Henares, Spain, October 3–4, 2024
Proceedings

Editors
Sanju Tiwari
Sharda University
Noida, Uttar Pradesh, India

Shodhguru
New Delhi, India

Miguel-Angel Sicilia
Universidad de Alcalá de Henares
Alcala de Henares, Spain

Fernando Ortiz-Rodriguez
Universidad Autónoma de Tamaulipas
Tamaulipas, Mexico

Tek Raj Chhetri
Massachusetts Institute of Technology (MIT)
Boston, MA, USA

ISSN 1865-0929 ISSN 1865-0937 (electronic)
Communications in Computer and Information Science
ISBN 978-3-031-81368-9 ISBN 978-3-031-81369-6 (eBook)
https://doi.org/10.1007/978-3-031-81369-6

This Springer imprint is published by the registered company Springer Nature Switzerland AG
The registered company address is: Gewerbestrasse 11, 6330 Cham, Switzerland

Preface

Artificial Intelligence (AI) is a term used frequently nowadays, and its impact on different areas is increasing. AI is expected to positively affect sustainability and the achievement of sustainable development goals (SDGs). However, to date, there are very few published studies systematically assessing the extent to which AI might impact all aspects of sustainable development. Given the speed with which artificial intelligence is evolving and its potential to disrupt many sectors, with advancements in Big Data, Hardware, Semantic Web, Machine Learning, Smart cities technologies, weather forecasting, and emerging powerful AI Algorithms, it seems that all the pieces are coming together to make huge changes to our everyday lives. Improving the planet Earth doesn't seem as hard as it used to because of these advancements.

AI holds great promise for building inclusive knowledge societies and helping countries reach their targets under the 2030 Agenda for Sustainable Development, but it also poses acute ethical challenges. The current choices to develop a sustainable-development-friendly AI by 2030 have the potential to unlock benefits that could go far beyond the SDGs within our century.

The conference Artificial Intelligence: Towards Sustainable Intelligence (AI4S) aims to open a discussion on trustworthy AI and related topics, and to present the most up-to-date developments around the world from researchers and practitioners. The main scientific program of the conference comprised 18 papers: 16 full research papers and 2 short research papers selected out of 59 reviewed submissions, which corresponds to an acceptance rate of 30%.

The General and Program Committee chairs would like to thank the many people involved in making AI4S 2024 a success. First, our thanks go to the main chairs and the 62 reviewers for ensuring a rigorous review process that, with an average of three double-blind reviews per paper, led to an excellent scientific program. The whole AI4S whole team is grateful to esteemed keynote speakers Ahmet Soylu, Kristiania University College, Oslo, Norway, and Prasad Yalamanchi, CEO at Lead Semantics, USA for their wonderful sessions. Special thanks to the University of Alcalá and particularly to Miguel Angel Sicilia for hosting the event AI4S in Alcalá de Henares, Spain.

Further, we are thankful for the kind support of the team at Springer. We finally thank our sponsors for their vital support of AI4S 2024. The editors would like to close the preface with warm thanks to our supporting keynotes and our enthusiastic authors who made this event truly international.

October 2024

Sanju Tiwari
Fernando Ortiz-Rodriguez
Miguel Angel Sicilia
Tek Raj Chhetri

Organization

General Chairs

Sanju Tiwari	ShodhGuru Innovation and Research Labs, India
Fernando Ortiz-Rodríguez	Autónoma de Tamaulipas, Mexico
Miguel-Angel Sicilia	University of Alcalá, Spain
Tek Raj Chhetri	Massachusetts Institute of Technology, USA

Program Chairs

Sashikala Mishra	Symbiosis Institute of Technology, India
Manas Gaur	UMBC, USA
Sven Groppe	University of Lübeck, Germany

Local Chairs

Marcal Mora-Cantallops	University of Alcalá, Spain
Lino González García	University of Alcalá, Spain
Alberto Ballesteros Rodríguez	University of Alcalá, Spain

Publicity Chairs

Ronak Panchal	Cognizant, India
Fatima Zahra Amara	University of Khenchela, Algeria
Ellie Young	Common Action, USA
Vania V. Estrela	Universidade Federal Fluminense, Brazil
Patience Usoro Usip	University of Uyo, Nigeria
Shishir Shandilya	VIT Bhopal University, India
M. A. Jabbar	Vardhman Engineering College, India

Tutorial Chairs

Sonali Vyas	UPES, India
Kusum Lata	Sharda University, India

Rajan Gupta	Analyttica Datalab, India
Valentina Janev	University of Belgrade, Serbia
Rita Zhgeib	Canadian University Dubai, UAE

Workshop Chairs

Namrata Nagpal	Amity University, India
Meenakshi Srivastava	Amity University, India
Ghanpriya Singh	National Institute of Technology, Kurukshetra, India

Special Session Chairs

Amna Dridi	Birmingham City University, UK
Janneth Alexandra Chicaiza Espinosa	Universidad Técnica Particular de Loja, Ecuador
Sailesh Iyer	Rai University, India
Soror Sahri	Université Paris Descartes, France

Program Committee

Alycia Sebastian	Al Zahra College for Women, Oman
Amna Dridi	Birmingham City University, UK
Amed Leiva-Mederos	Central University of Las Villas, Cuba
Ashwani Kumar Dubey	Amity University, Noida, India
Beyzanur Çayır Ervural	Necmettin Erbakan University, Turkey
Carlos F. Enguix	Universidad Autónoma de Tamaulipas, Mexico
Daniel E. Asuquo	University of Uyo, Nigeria
David Martín-Moncunill	Universidad Camilo José Cela, Spain
Deepali Vora	Symbiosis Institute of Technology, India
Edgar Tello Leal	Universidad Autónoma de Tamaulipas, Mexico
Edward Udo	University of Uyo, Nigeria
Ekin Akkol	İzmir Bakırçay University
Fatima Zahra Amara	University of Khenchela, Algeria
Fernando Ortiz-Rodríguez	Universidad Autónoma de Tamaulipas, Mexico
Gerard Deepak	Manipal Institute of Technology Bengaluru, India
Gerardo Haces	Universidad Autónoma de Tamaulipas, Mexico
Ghanpriya Singh	NIT Kurukshetra, India
Gustavo de Asis Costa	Federal Institute of Education, Brazil
Hugo Eduardo Camacho Cruz	Universidad Autónoma de Tamaulipas, Mexico

Ifiok James Udo	University of Uyo, Nigeria
Jose L. Martinez-Rodriguez	Universidad Autónoma de Tamaulipas, Mexico
Jose Melchor Medina-Quintero	Universidad Autónoma de Tamaulipas, Mexico
Jude Hemanth	Karunya University, India
Jannathl Firdouse Mohamed Kasim	Al Zahra College for Women, Oman
Jitendra Kumar Samriya	IIIT Ropar, India
Janneth Alexandra Chicaiza Espinosa	UTP de Loja, Ecuador
Kusum Lata	Sharda University, India
M. A. Jabbar	Vardhman Engineering College, India
Marcin Paprzycki	Polish Academy of Sciences, Poland
Meenakshi Srivastava	Amity University, Lucknow, India
Meriem Djezzar	Khenchela University, Algeria
Miguel-Angel Sicilia	University of Alcalá, Spain
Mounir Hemam	Khenchela University, Algeria
Müge Oluçoğlu	İzmir Bakırçay Üniversitesi
Namrata Nagpal	Amity University, Lucknow, India
Narottam Das Patel	VIT Bhopal, India
Ömer Faruk Yılmaz	Karadeniz Technical University, Turkey
Onur Dogan	University of Padua, Italy
Ourania Areta Hiziroğlu	İzmir Bakırçay University, Turkey
Piyush Kumar Pareek	NMIT, India
Prateek Thakral	JUIT, India
Patience Usoro Usip	University of Uyo, Nigeria
Ronak Panchal	Cognizant, India
Ritu Tanwar	NIT Uttarakhand, India
Sachinandan Mohany	VIT-AP University, India
Sanju Tiwari	Universidad Autónoma de Tamaulipas, Mexico
Sailesh Iyer	Rai University, India
Sarra Ben Abbes	Gireve, France
Sashikala Mishra	Symbiosis Institute of Technology, India
Sonali Vyas	UPES, India
Seema Verma	Delhi Technical Campus, India
Serge Sonfack	INP-Toulouse, France
Shishir Shandilya	VIT Bhopal University, India
Sanchari Saha	CMR Institute of Technology, India
Syed Saba Raoof	Vardhaman College of Engineering, India
Valentina Janev	Institute Mihajlo Pupin, Serbia
Venkata Lakshmi Durga	Sarala Birla University, India
Vijaya Padmanabha	Modern College of Business and Science, Oman
Yevheniia Znakovska	National Aviation University, Ukraine

Contents

Sustainability Performance Through a Business Process Mining Based Conceptual Framework Integrating GRI Metrics

Ourania Areta Hiziroglu(✉) and Onur Dogan

Department of Management Information Systems, Izmir Bakircay University, 35665 Izmir, Turkey
{ourania.areta,onur.dogan}@bakircay.edu.tr

Abstract. In the global pursuit of sustainable business practices, organizations face the challenge of effectively measuring and enhancing their sustainability performance. Process mining has emerged as a pivotal tool in this endeavour, offering data-driven insights that enable the analysis and visualization of operational processes. This approach uncovers inefficiencies, bottlenecks, and deviations from intended workflows, facilitating the identification of key sustainability metrics such as resource utilization, energy consumption, and waste generation. However, despite the increasing pressure to adopt sustainable practices, many sectors lack comprehensive methods to assess and optimize sustainability across complex operational processes. This study addresses this gap by proposing a novel conceptual framework that leverages business process mining techniques to measure and improve sustainability performance, specifically incorporating Global Reporting Initiative (GRI) metrics. Our framework integrates process mining with GRI sustainability indicators to identify inefficiencies, bottlenecks, and improvement opportunities throughout the value chain. We present a detailed model that maps GRI metrics to specific process steps, enabling a standardized yet flexible approach to sustainability performance measurement. The proposed framework demonstrates the potential of process mining as a powerful tool for driving sustainable practices, offering a new approach to tackling pressing environmental challenges while aligning with globally recognized reporting standards. This framework provides a continuous monitoring and improvement methodology for organizations to improve their sustainability performance, identify improvement opportunities, and provide transparent reporting. It bridges the gap between high-level sustainability reporting and operational processes, promoting a culture of continuous improvement across diverse industries and operational environments.

Keywords: Sustainability · Process mining · GRI metrics

S. Tiwari et al. (Eds.): AI4S 2024, CCIS 2243, pp. 1–15, 2025.
https://doi.org/10.1007/978-3-031-81369-6_1

1 Introduction

In the last few decades, sustainability has become an increasing concern for corporations due to rising ecological problems, social expectations, and legal obligations [29]. Corporate sustainability has slowly shifted from a peripheral consideration to a core strategic management concern in organizations cutting across industries and sectors [7]. Sustainability issues are complex and interrelated in most business ventures. In the environmental domain, management faces problems such as global warming, depletion of natural resources, and loss of biological diversity. The need to decrease the levels of emissions and waste while maximizing the circularity of products and services has been brought to the forefront, especially given the goals established under the 2015 Paris Climate Accord [4] [14]. This includes issues related to labor standards, diversity and inclusion, community well-being, and human rights within supply chains. COVID-19 has amplified the role of social sustainability in the context of organizational endurance and inclusion and socially responsible business models [27]. Economically, an organization must achieve current financial performance to meet the present need and consider future sustainability while harnessing opportunities in new, sustainable markets [30]. Even with the availability of sustainability reporting frameworks such as the Global Reporting Initiative (GRI), companies have an issue implementing measurements for sustainability performance due to challenges in converting broad, abstract objectives into practical, tangible operations [15].

Process mining has been increasingly recognized for enhancing sustainability management within organizations. Process mining facilitates the extraction and analysis of event logs generated by information systems by employing data-driven techniques. This methodology enables a thorough understanding of actual process execution, identifying inefficiencies, bottlenecks, and deviations from the intended workflows. As a result, operational inefficiencies that contribute to unnecessary resource consumption and waste generation are highlighted.

In the context of sustainability management, process mining serves as a critical tool for aligning operational processes with sustainability goals. Integrating process mining with sustainability metrics allows for precise measuring of key indicators such as resource utilization, energy consumption, and emissions. These insights are instrumental in pinpointing areas where sustainability practices can be enhanced, thus driving more effective resource management and reducing the environmental footprint of organizational operations.

Moreover, process mining facilitates the continuous monitoring and improvement of sustainability performance. Providing a dynamic and real-time view of processes enables organizations to track progress against sustainability targets and quickly respond to deviations. This proactive approach ensures that sustainability initiatives are not only implemented but also sustained over time, fostering a culture of ongoing improvement.

The application of process mining in sustainability management also supports transparency and accountability. Organizations can generate detailed and accurate reports that align with globally recognized standards, such as GRI, by mapping sustainability metrics to specific process steps. This alignment enhances the credibility of sustainability reporting and helps build trust with stakeholders by demonstrating a commitment to transparent and responsible business practices.

This study addresses the gap between sustainability aspirations and actual performance by proposing a novel conceptual framework that leverages process mining techniques to measure and improve sustainability performance, specifically incorporating GRI metrics. The objectives of this research are the development of a conceptual framework that takes advantage of process mining tools for sustainability performance measurement and the integration of GRI metrics into this process mining approach.

This paper enhances the growing body of literature on sustainable business practices by tackling these objectives. It offers a practical tool for organizations looking to improve their sustainability performance [10]. The framework of the study provides a way to transform high-level sustainability goals into tangible, measurable improvements in business processes, addressing a critical need in the field of corporate sustainability management.

2 Literature Review

2.1 Sustainability Performance Measurement

Sustainability performance measurement is a vital part of organizational sustainability since it enables the assessment of organizational activities' environmental, social, and economic effects. The literature shows a shift towards the construction of integrated frameworks embodying all these three sustainability dimensions, often referred to as the Triple Bottom Line (TBL) [28].

A study of the UK manufacturing sector conducted empirical research that uncovered the intricate correlation between sustainable supply chain management and organizational success. Although sustainable procurement positively impacted economic performance, implementing sustainable supply chain management did not always result in improved financial outcomes. This highlights the importance of nuanced approaches in measuring sustainability performance [8].

Many current sustainability measuring frameworks primarily emphasize environmental aspects. This finding underscores the significance of embracing a comprehensive strategy that gives equal consideration to both social and economic elements [17]. An extensive literature study has found measures connected to social concerns in sustainable supply chains. This research highlights the increasing awareness of social variables in measuring sustainability performance [2].

Standardized measurements and decision-support systems are essential for promoting sustainable industry practices. This need is especially clear when producing next-generation biofuels, where it is crucial to deal with uncertainties related to environmental and sustainability considerations [34].

Other work has suggested a comprehensive framework for measuring a product's sustainability performance. This framework includes concepts, such as distinguishing between resource and value metrics, clearly representing the TBL, and considering the Life Cycle Assessment (LCA). This paradigm represents the shift towards more inclusive and interconnected methods of measuring sustainability performance [28].

Recent research has also concentrated on quantitatively assessing sustainability performance in small and medium-sized companies (SMEs) by acknowledging the distinct difficulties encountered by various kinds of organizations. Research has emphasized the need to develop customized frameworks suitable for the particular circumstances and limitations of SMEs [18].

Moreover, the industry has further comprehended the need for monitoring sustainability performance by developing particular economic, environmental, and social measures [20]. Recent studies have stressed the importance of sustainability performance metrics as markers of effective sustainability initiatives inside enterprises. They also highlight the significance of performance assessment in promoting and verifying sustainability efforts [11].

Furthermore, several studies have used many methodologies (and conceptual frameworks) to evaluate sustainability performance. The presence of many types of metrics underscores the significance of choosing suitable measurements that align with the business's objectives and the unique issues faced by the sector [25].

Ultimately, assessing sustainability performance is now undergoing significant and continuous development. There is a noticeable shift towards adopting holistic strategies that include environmental, social, and economic aspects in a balanced manner. A crucial problem arises in creating standardized measurements adaptable to particular organizational settings and industries. Potential areas for future study are improving industry-specific measurements, strengthening the incorporation of social variables, and making more advanced methodologies for gathering and analyzing data to assess sustainability performance.

2.2 Global Reporting Initiative (GRI) Metrics

The significance of corporate sustainability performance measurement systems has been emphasized, stressing the importance of structured research questions to guide future work in this field [31]. A benchmark for comprehensive sustainability reporting practices has been established by integrating sustainability indicators into performance measurement systems, as spearheaded GRI [21]. According to the GRI [13] "The GRI Standards enable an organization to publicly disclose its most significant impacts on the economy, environment, and people, including impacts on their human rights and how the organization manages these impacts. This enhances transparency on the organization's impacts and increases organizational accountability".

Studies on the method and tools used in implementing sustainability and assessing sustainability also pointed out the need to engage in frameworks such as the GRI for sustainability reporting [3]. It has been dramatically argued that

GRI metrics have critically shaped the standardization of sustainability reporting practices, enabling organizations to achieve sustainable performance that is reportable consistently and transparently [19]. The GRI framework has played an important role and has provided insights for policy implications and strategic decision-making through its connection between sustainability indicators and the Sustainable Development Goals (SDGs) [19].

The significance of comprehensive literature reviews in conceptualizing performance indicators was highlighted in a sustainability performance measurement framework formulated for projects in the UK [35]. Different models of evaluating sustainability performance have adopted GRI measures, which offer an organized model to organizations to evaluate their sustainability efforts [1]. Sustainability advancements and sustainable organizational development that lead to better environmental, social, and economically sustainable performance are achievable through integrating operational programs for sustainable performance goals [6].

Promoting sustainability reporting practices and encouraging accountability and transparency have been vital benefits of utilizing GRI metrics within the performance measurement framework. To measure and evaluate how corporations perform on the sustainability agenda and demonstrate commitment to sustainable business practices, corporate sustainability performance measurement systems should integrate with key performance indicators illustrating performance on GRI principles [32]. With the help of GRI metrics through the provision of a perceptive framework for assessing and tracking sustainability performance, the organizations have been enabled not only to create positive environmental/social impact but also to pursue and achieve sustainable business success.

Therefore, with the rise of the GRI metrics, sustainability reporting practices have been revolutionized since organizations now have a definite benchmark that measures their environmentally, socially, and economically sustainable performance. The GRI framework has made transparency and responsible approaches to sustainable practices possible through reporting activities, striving to adhere to international standards and norms. Measures in the form of GRI indicators that can improve sustainability programs, drive operational effectiveness, and support a more sustainable environment in companies can be integrated into organizational performance measurement frameworks.

2.3 Process Mining and Sustainability

The incorporation of process mining techniques and sustainability measures is a developing area of study that has the potential to greatly improve the sustainability performance of organizations. Contemporary research on process mining addresses sustainability concepts only to a limited extent and often implicitly.

To start with, Horsthofer-Rauch et al. [16] suggest a new method that merges sustainability-integrated value stream mapping with process mining. Their paradigm thoroughly examines and improves value streams in industrial processes while also considering sustainability factors. This research showcases the practicality of combining process mining approaches with sustainability measures in real-world situations by offering implementation details and confirming the concept via practical case studies.

Then, the study by Graves et al. [12] emphasizes a significant deficiency in the utilization of process mining in sustainability. Their study demonstrates that process mining has considerable promise for evaluating and analyzing sustainability in corporate processes, but its utilization in this domain is still restricted. To fill this need, the authors provide the PM4S framework, which promotes collaboration between the process mining and sustainability communities. This framework signifies a significant advancement in integrating these two disciplines and using process mining to enhance sustainability.

Moreover, Safitri et al. [26] investigate the application of process mining in assessing company sustainability indicators. Their research showcases using sustainability metrics to classify business processes, resulting in enhanced overall business performance. This study offers a realistic method for incorporating sustainability criteria into process analysis and development endeavors by utilizing event logs from process mining as a benchmark.

Furthermore, some studies explore the environmental impact of business processes concerning CO2 emissions [5]. Other research applies process mining to examine social factors such as trust, privacy issues, customer satisfaction, fairness, and economic aspects [23,24,33]. Additionally, a branch of research focuses on Green Business Process Management, which outlines metrics for sustainable processes without employing process mining for deeper analysis [26]. Lastly, Ortmeier et al. [22] proposed a framework integrating process mining into Life Cycle Assessment (LCA). This method involves incorporating energy and resource data into the event log and visualizing energy consumption within the process map.

These studies show an increasing recognition of the possible synergies between process mining and sustainability measurement. Their merging provides numerous significant advantages. Process mining approaches may give extensive, data-driven insights into corporate processes, allowing for better-informed decision-making on sustainability changes. Organizations may improve their sustainability performance by mapping sustainability measures to particular process stages. The dynamic nature of process mining enables constant monitoring of sustainability performance, fostering a culture of continuous improvement. Integrating sustainability indicators into process mining can result in more open reporting and accountability for sustainability performance.

2.4 Gaps in Current Sustainability Assessment Methods

However, a literature review has also identified obstacles, gaps, and opportunities for future inquiry. More specifically, while current sustainability assessment approaches are advancing, some critical gaps exist in successfully assessing and improving organizational sustainability performance. These shortcomings are especially noticeable when examining the possible integration of process mining approaches with sustainability criteria.

First, there is a noticeable lack of dynamic and real-time sustainability performance assessment. Most existing methodologies rely on static, periodic evaluations, which may fail to capture daily fluctuations in sustainability performance. Horsthofer-Rauch et al. [16] emphasize this gap by proposing integrating sustainability measures into value stream mapping, indicating the necessity for more continuous and process-oriented sustainability evaluation.

Second, existing sustainability evaluation approaches lack adequate granularity. Many techniques concentrate on high-level, organizational-wide measures, neglecting to give information on how individual business activities contribute to overall sustainability performance. While studies may have addressed this gap [26], nevertheless, more complete, process-level sustainability evaluation approaches are still needed.

Then, there is a lack of connection between sustainability assessments and operational procedures. According to Graves et al. [12], process mining has yet to be widely implemented in the sustainability sector, indicating a considerable gap in the use of operational data for sustainability evaluation and improvement.

Fourth, present approaches frequently fail to balance and integrate various aspects of sustainability. While environmental indicators are often well-developed, sustainability's social and economic components may be overlooked or inadequately incorporated into overall evaluation frameworks.

Fifth, present techniques cannot frequently give practical feedback for change. While they may indicate areas of poor sustainability performance, they may not provide apparent solutions for improving these areas. The work of Gallotta et al. [9] in modeling scenarios for attaining sustainability in corporate processes highlights this gap and the need for more practical, implementation-oriented approaches.

Finally, sustainability evaluation methodologies tailored to individual industries are lacking. Many existing techniques are broad and may not sufficiently represent the distinct sustainability issues and possibilities in different sectors.

These shortcomings underscore the importance of more integrated, dynamic, and process-oriented methods for sustainability evaluation. The possible integration of process mining methodologies with sustainability measurements opens up new options for resolving these gaps, allowing for more effective assessment, monitoring, and improvement of corporate sustainability performance.

However, standardized methods are needed to combine sustainability measures with process mining methodologies. To fully realize the benefits of this integration, process mining specialists and sustainability professionals must work together more closely. Further research is needed to determine how this integrated strategy might be used in various industry settings, and to validate the efficacy of integrated methods at various organizational sizes and conditions.

These problems highlight the complexities of combining process mining with sustainability indicators, emphasizing the importance of ongoing study and development in this new discipline.

3 Conceptual Framework

Considering those mentioned above, this study has developed and proposes a conceptual framework illustrated in Fig. 1. The proposed framework integrates GRI metrics with business process mining techniques to enhance organizational sustainability performance. It emphasizes the comprehensive capture and analysis of sustainability-related data from various perspectives, including environmental, social, and economic metrics. By leveraging GRI metrics, the framework ensures that sustainability performance is measured against widely recognized standards.

The combination of process mining and GRI metrics helps to achieve a better understanding and higher granularity of the changes in sustainability performance over time [6]. Using key process maps for GRI indicators, it is possible to determine the areas of increased sustainability risks and potential for improvement at the operational level.

The framework considers and integrates two essential perspectives: sustainability and process mining. From the sustainability perspective, different key performance indicators (KPIs) are categorized under environmental (En-KPI), social (S-KPI), and economic (Ec-KPI) categories. These metrics are gathered and stored in a GRI metrics repository. Simultaneously, an event log capturing the organization's operational activities is maintained. The event log and the GRI metrics are combined to create an enriched event log, which integrates sustainability metrics with operational process data. This enriched event log forms the basis for applying various process mining techniques.

3.1 Sustainability Perspective

The sustainability perspective in our proposed framework is built upon the comprehensive integration of GRI metrics into the process mining approach. This perspective encompasses multiple dimensions of sustainability, providing a holistic view of an organization's performance:

1. Environmental Metrics (En-KPI): These indicators concern how an organization affects living and nonliving natural resources-ecosystems, land, air, and water. En-KPIs include energy use, $CO2$ emissions, water use, and waste disposal.
2. Social Metrics (S-KPI): These metrics focus on the effects an organization causes within the social systems in which it is located. S-KPIs could involve labour relations, human rights, product safety, and social impact on communities.
3. Economic Metrics (Ec-KPI): They quantify the economic effects of the organization at the national and international levels. The Ec-KPIs may encompass measures of financial performance, market presence, indirect economic impacts, and procurement practices.

The sustainability approach is implemented through the following components:

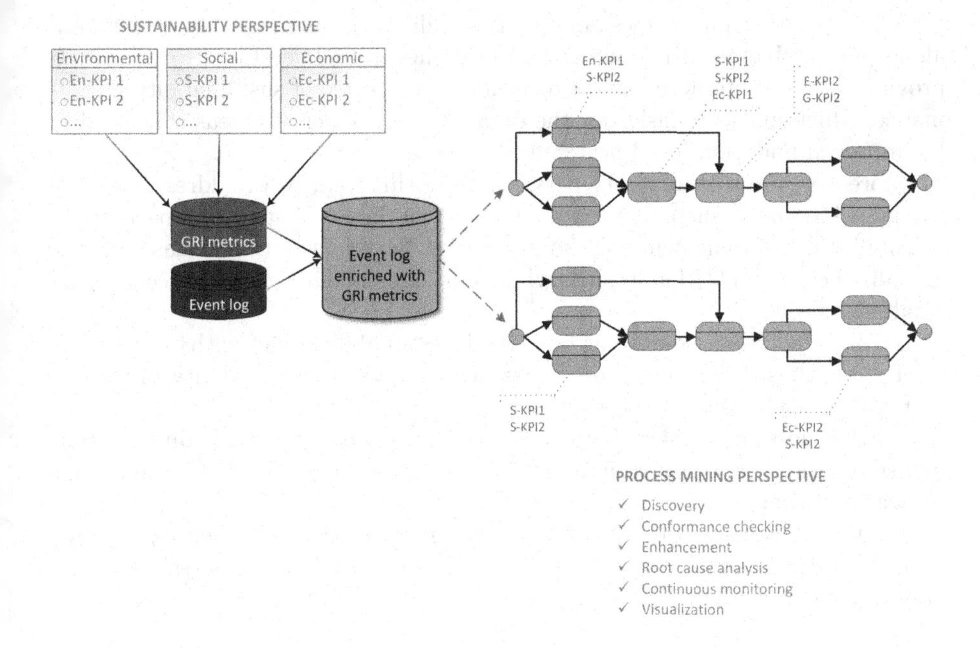

Fig. 1. Conceptual Framework for Integrating Sustainability Perspective with Process Mining using GRI Metrics

- GRI Metrics Repository: A comprehensive database of standardized sustainability metrics based on GRI guidelines. It acts as the main repository for the sustainability performance measurement throughout the organization. The depository includes all metrics that are included under each category of the standards and meet TBL sustainability aspects:
  - Environmental Metrics (GRI 300 series): Approximately 32 specific disclosures across eight environmental standards.
  - Social Metrics (GRI 400 series): Approximately 40 specific disclosures across 19 social standards.
  - Economic Metrics (GRI 200 series): Approximately 13 specific disclosures across 7 economic standards [13]. Companies may choose the most relevant metrics for their operations and industry.
- Metric Mapping: GRI metrics are systematically mapped to specific process steps within the organization. This mapping enables a granular analysis of sustainability performance at the operational level. This method involves identifying key GRI indicators relevant to the organization and aligning them with corresponding business activities. For example, metrics related to energy consumption, waste generation, and resource utilization can be mapped to specific manufacturing processes or supply chain activities.
- Enriched Event Log: The framework links operational event logs with GRI metrics, forming an augmented dataset incorporating sustainability performance data into process data.

This approach guarantees that sustainability measurement is not a standalone process but is rather integrated into the firm's operational practices. It provides a more interactive and executable perspective of sustainability performance, which makes it easier for the organization to identify areas that need to be improved and monitored over time.

Moreover, the sustainability perspective in this framework addresses several key gaps in current sustainability assessment methods. First, it approaches sustainability measurement in a standardized yet flexible way that aligns with the globally recognized GRI standards while allowing customization to specific organizational needs.

Then, it facilitates combining high-level sustainability goals with operational-level data, thus decreasing the gap between strategic sustainability objectives and day-to-day business processes.

Furthermore, it enables the monitoring and review of sustainability performance by using data on sustainability indicators at the process level in real time or near real time.

Finally, it considers environmental, social, and economic factors, aligning with the TBL approach to sustainability and offering a holistic assessment of an organization's sustainability performance.

3.2 Process Mining Perspective

By integrating these process mining techniques with GRI sustainability metrics, the framework enables comprehensive measurement and facilitates continuous improvement in sustainability practices across operational processes. It aligns operational details with high-level sustainability reporting standards, promoting transparency and accountability in sustainable business practices. Several process mining techniques can be effectively utilized in the context of sustainability performance. Figure 1 depicts how GRI metrics are combined with process mining techniques.

Process discovery techniques involve extracting process models from event logs. It means capturing and analyzing data from operational processes related to sustainability metrics like resource utilization, energy consumption, and waste generation. Process mining discovery algorithms can automatically discover process flows, showing how activities are interconnected and where inefficiencies may lie.

Conformance checking techniques compare discovered process models with predefined models or rules. It helps assess whether actual processes adhere to sustainability standards and GRI metrics. Deviations can indicate areas where sustainability goals are not being met or where processes can be optimized.

Process mining allows for enhancing existing process models based on actual data. This involves improving process performance and sustainability by identifying and implementing best practices, such as optimizing resource use or reducing waste based on insights derived from process mining analysis.

Root cause analysis can be applied to understand the main reasons for GRI problems, which are crucial. Process mining can help trace back to the origins of inefficiencies or deviations from sustainability metrics, such as identifying specific process steps or conditions that lead to excessive energy consumption or waste generation.

Process mining facilitates continuous monitoring of sustainability performance metrics. Real-time or periodic analysis can provide feedback on progress toward sustainability goals, enabling timely interventions and adjustments.

Visual representations of process flows and metrics are integral to understanding and communicating sustainability performance. Process mining often provides visual dashboards and charts highlighting KPIs and improvement areas, making data-driven decision-making more accessible. By visualizing how each process step contributes to overall sustainability goals, companies can implement targeted interventions to enhance their environmental and social impact. Additionally, this approach supports continuous improvement by providing real-time data on process performance and facilitating timely adjustments and optimizations.

4 Discussion and Conclusion

The integration of process mining techniques with GRI metrics is essential for the proposed conceptual framework. The framework is a novel contribution to the field of sustainability performance measurement and management. The presented approach fills several significant voids in current sustainability assessment approaches and brings many advantages for organizations seeking to improve their sustainability profiles.

Another advantage is the fact that this framework can be used to generate an effective real time and dynamic measure of sustainability performance. This approach differs from regular assessment techniques such as checklists, gap analysis, and scorecards because it can monitor sustainability performance indicators in live business processes through process mining techniques. This real-time capacity ensures that organizations can easily detect and address sustainability concerns as they emerge, making sustainability management more proactive.

Furthermore, the framework takes a more detailed approach to assessment regarding sustainability. It is also important to note that by linking GRI metrics to process steps, an organization can track and understand the contribution of particular business processes about the sustainability picture. This contradicts current sustainability assessment methods, where most provide simplified, general organization sustainability performance indicators. The more detailed approach allows for better identification of problems and potential enhancements, potentially resulting in better sustainability practices being implemented.

Reflecting sustainability assessment into the operational processes is an area that has not been well explored in operations. Regarding sustainability, Graves et al. [12] pointed out that process mining has not advanced to be adopted massively. This framework serves as a bridge to close that gap, as it uses operational data for sustainability assessment and enhancement. Thus, it connects sustainability issues to everyday business practices and actions, meaning that sustainable development goals are not an additional factor but an inherent part of organizational processes.

The framework's holistic approach, which incorporates environmental, social, and economic measurements, is consistent with the TBL notion of sustainability. This holistic approach guarantees that companies consider all aspects of sustainability while assessing and improving. It solves the prevalent issue of prioritizing environmental measures above social and economic variables, resulting in a more balanced and comprehensive view of sustainability performance.

Moreover, the approach this study proposes can provide practical input for improvement. Organizations may utilize process mining approaches in conjunction with sustainability indicators to not only identify areas of poor sustainability performance but also analyze the underlying process-related issues. This greater knowledge can result in more effective and focused improvement attempts.

The framework's adaptability is essential. On one hand, it includes conventional criteria (GRI), and on the other, it enables customisation, which can be derived from organizational needs and industry peculiarities. This type of flexibility is necessary considering that there are a wide variety of sustainability concerns and issues among industries and organizations.

However, the usage of the framework may create several issues. One possible impediment is the requirement for high-quality, comprehensive data. The efficacy of process mining approaches strongly depends on the availability and quality of event logs and other operational data. Organizations may need to invest in enhanced data gathering and administration tools to benefit from it fully.

Another problem is integrating sustainability expertise with process mining understanding. The practical application of this approach would need collaboration between sustainability professionals and process mining experts, two groups that may not have previously worked closely together. Organizations must encourage cross-disciplinary cooperation to get the most benefits from the framework.

The framework's complexity may also make it difficult, especially for smaller firms or those with limited resources. Implementing and maintaining a comprehensive system for measuring sustainability performance may require a large investment of time, knowledge, and technology.

Despite these obstacles, this paradigm has significant potential advantages. It has the potential to significantly enhance corporate sustainability practices by offering a more dynamic, granular, and integrated approach to sustainability performance monitoring.

This conceptual framework represents a substantial advancement in sustainability performance assessment and management. Its capacity to give real-time insights, detailed approach to sustainability evaluation, and integration of sustainability issues into operational processes are remarkable. These elements can change how firms approach sustainability, turning it into an integrated aspect of their operations rather than a separate, compliance-driven activity.

While there are implementation obstacles, notably in data needs and the need for cross-disciplinary competence, the framework can potentially provide significant advantages. It provides businesses a valuable tool for improving sustainability performance, promoting continuous improvement, and meeting sustainability objectives.

Future studies should focus on empirical validation of this paradigm in various sectors and organizational settings. Furthermore, more effort is required to produce practical implementation recommendations, especially for smaller businesses or those with fewer resources. As sustainability evolves, frameworks like this one will be critical in assisting companies in navigating the challenging environment of measuring and improving sustainable performance.

This conceptual framework can address the critical need for companies to monitor and manage sustainability performance more effectively. Harnessing the power of process mining and defined GRI indicators provides a roadmap to more sustainable and responsible business operations in an increasingly complex and challenging business environment.

References

1. Aalst, W., et al.: Business process mining: an industrial application. Inf. Syst. **32**(5), 713–732 (2007). https://doi.org/10.1016/j.is.2006.05.003
2. Ahi, P., Searcy, C.: Measuring social issues in sustainable supply chains. Meas. Bus. Excell. **19**(1), 33–45 (2015). https://doi.org/10.1108/mbe-11-2014-0041
3. Amaral, L., Martins, N., Gouveia, J.: Quest for a sustainable university: a review. Int. J. Sustain. High. Educ. **16**(2), 155–172 (2015). https://doi.org/10.1108/ijshe-02-2013-0017
4. Baumgartner, R.J., Rauter, R.: Strategic perspectives of corporate sustainability management to develop a sustainable organization. J. Clean. Prod. **140**, 81–92 (2017). https://doi.org/10.1016/j.jclepro.2016.04.146
5. Brehm, L., Slamka, J., Nickmann, A.: Process mining for carbon accounting: an analysis of requirements and potentials. In: Digitalization Across Organizational Levels: New Frontiers for Information Systems Research, pp. 209–244. Springer (2022) https://doi.org/10.1007/978-3-031-06543-9_9
6. Caiado, R., Quelhas, O., Nascimento, D., Anholon, R., Filho, W.: Towards sustainability by aligning operational programmes and sustainable performance measures. Prod. Plann. Control **30**(5–6), 413–425 (2019). https://doi.org/10.1080/09537287.2018.1501817
7. Engert, S., Rauter, R., Baumgartner, R.J.: Exploring the integration of corporate sustainability into strategic management: a literature review. J. Clean. Prod. **112**, 2833–2850 (2016). https://doi.org/10.1016/j.jclepro.2015.08.031
8. Esfahbodi, A., Zhang, Y., Watson, G., Zhang, T.: Governance pressures and performance outcomes of sustainable supply chain management - an empirical analysis of uk manufacturing industry. J. Clean. Prod. **155**, 66–78 (2017). https://doi.org/10.1016/j.jclepro.2016.07.098
9. Gallotta, B., Garza-Reyes, J., Anosike, A., Lim, M., Roberts, I.: A conceptual framework for the implementation of sustainability business processes. In: Production and Operations Management Society (POMS) (2016)
10. Gao, J., Bansal, P.: Instrumental and integrative logics in business sustainability. J. Bus. Ethics **112**(2), 241–255 (2012). https://doi.org/10.1007/s10551-012-1245-2

11. Geyi, D., Yusuf, Y., Menhat, M., Abubakar, T., Ogbuke, N.: Agile capabilities as necessary conditions for maximising sustainable supply chain performance: an empirical investigation. Int. J. Prod. Econ. **222**, 107501 (2020). https://doi.org/10.1016/j.ijpe.2019.09.022
12. Graves, N., Koren, I., van der Aalst, W.M.: Rethink your processes! a review of process mining for sustainability. In: 2023 International Conference on ICT for Sustainability (ICT4S), pp. 164–175 (2023).https://doi.org/10.1109/ICT4S58814.2023.00025
13. GRI: Gri (2024). https://www.globalreporting.org/. Accessed 17 July 2024
14. Haffar, M., Searcy, C.: Target-setting for ecological resilience: are companies setting environmental sustainability targets in line with planetary thresholds? Bus. Strateg. Environ. **27**(7), 1079–1092 (2018). https://doi.org/10.1002/bse.2053
15. Hahn, R., Kühnen, M.: Determinants of sustainability reporting: a review of results, trends, theory, and opportunities in an expanding field of research. J. Clean. Prod. **59**, 5–21 (2013). https://doi.org/10.1016/j.jclepro.2013.07.005
16. Horsthofer-Rauch, J., et al.: Sustainability-integrated value stream mapping with process mining. Prod. Manuf. Res. **12**(1), 2334294 (2024). https://doi.org/10.1080/21693277.2024.2334294
17. Labuschagne, C., Brent, A., Erck, R.: Assessing the sustainability performances of industries. J. Clean. Prod. **13**(4), 373–385 (2005). https://doi.org/10.1016/j.jclepro.2003.10.007
18. Malesios, C., De, D., Moursellas, A., Dey, P., Evangelinos, K.: Sustainability performance analysis of small and medium sized enterprises: criteria, methods and framework. Socioecon. Plann. Sci. **75**, 100993 (2021). https://doi.org/10.1016/j.seps.2020.100993
19. Mengistu, A., Panizzolo, R.: Tailoring sustainability indicators to small and medium enterprises for measuring industrial sustainability performance. Meas. Bus. Excell. **27**(1), 54–70 (2022). https://doi.org/10.1108/mbe-10-2021-0126
20. Mengistu, A., Panizzolo, R.: Metrics for measuring industrial sustainability performance in small and medium-sized enterprises. Int. J. Product. Perform. Manag. **73**(11), 46–68 (2023). https://doi.org/10.1108/ijppm-04-2022-0200
21. Nappi, V., Rozenfeld, H.: The incorporation of sustainability indicators into a performance measurement system. Procedia CIRP **26**, 7–12 (2015). https://doi.org/10.1016/j.procir.2014.07.114
22. Ortmeier, C., Henningsen, N., Langer, A., Reiswich, A., Karl, A., Herrmann, C.: Framework for the integration of process mining into life cycle assessment. Procedia CIRP **98**, 163–168 (2021). https://doi.org/10.1016/j.procir.2021.01.024
23. Pohl, T., Qafari, M.S., van der Aalst, W.M.: Discrimination-aware process mining: a discussion. In: International Conference on Process Mining, pp. 101–113. Springer (2022) https://doi.org/10.1007/978-3-031-27815-0_8
24. Qafari, M.S., Van der Aalst, W.: Fairness-aware process mining. In: On the Move to Meaningful Internet Systems: OTM 2019 Conferences: Confederated International Conferences: CoopIS, ODBASE, C&TC 2019, Rhodes, Greece, 21–25 October 2019, Proceedings, pp. 182–192. Springer (2019)https://doi.org/10.48550/arXiv.1908.11451
25. Ramani, T., Zietsman, J., Gudmundsson, H., Hall, R., Marsden, G.: Framework for sustainability assessment by transportation agencies. Transp. Res. Rec.: J. Transp. Res. Board **2242**(1), 9–18 (2011). https://doi.org/10.3141/2242-02
26. Safitri, L.N., Sarno, R., Budiawati, G.I.: Improving business process by evaluating enterprise sustainability indicators using fuzzy rule based classification. In: 2018

International Seminar on Application for Technology of Information and Communication, pp. 55–60. IEEE (2018).https://doi.org/10.1109/ISEMANTIC.2018.8549758
27. Sarkis, J.: Supply chain sustainability: learning from the COVID-19 pandemic. Int. J. Oper. Prod. Manage. **41**(1), 63–73 (2020). https://doi.org/10.1108/ijopm-08-2020-0568
28. Sartal, A., Rivera, R., Mejias, A., García-Collado, A.: The sustainable manufacturing concept, evolution and opportunities within industry 4.0: a literature review. Adv. Mech. Eng. **12**(5), 168781402092523 (2020). https://doi.org/10.1177/1687814020925232
29. Schaltegger, S., Hansen, E.G., Lüdeke-Freund, F.: Business models for sustainability. Organ. Environ. **29**(1), 3–10 (2015). https://doi.org/10.1177/1086026615599806
30. Schaltegger, S., Lüdeke-Freund, F., Hansen, E.G.: Business cases for sustainability and the role of business model innovation. centre for sustainability management (CSM) (2013). http://books.google.ie/books?id=DcfizwEACAAJ&dq=Business+cases+for+sustainability:+the+role+of+business+model+innovation+for+corporate+sustainability&hl=&cd=1&source=gbs_api
31. Searcy, C.: Corporate sustainability performance measurement systems: a review and research agenda. J. Bus. Ethics **107**(3), 239–253 (2011). https://doi.org/10.1007/s10551-011-1038-z
32. Searcy, C.: Updating corporate sustainability performance measurement systems. Meas. Bus. Excell. **15**(2), 44–56 (2011). https://doi.org/10.1108/13683041111131619
33. Werner, M.: Financial process mining-accounting data structure dependent control flow inference. Int. J. Account. Inf. Syst. **25**, 57–80 (2017). https://doi.org/10.1016/j.accinf.2017.03.004
34. Williams, P., Inman, D., Aden, A., Heath, G.: Environmental and sustainability factors associated with next-generation biofuels in the US.: what do we really know? Environ. Sci. Technol. **43**(13), 4763–4775 (2009). https://doi.org/10.1021/es900250d
35. Zhou, L., Keivani, R., Kurul, E.: Sustainability performance measurement framework for PFI projects in the UK. J. Financ. Manag. Prop. Constr. **18**(3), 232–250 (2013). https://doi.org/10.1108/jfmpc-08-2012-0032

KnowWhereGraph for Land Use Optimization: Achieving Sustainability and Efficiency

Michael McCain(✉), Rakesh Kandula, and Cogan Shimizu

Wright State University, Dayton, USA
{mccain.32,kandula.15,cogan.shimizu}@wright.edu

Abstract. This research aims to enhance land utilization for agricultural and renewable energy projects by optimizing the identification process of suitable locations. The challenge of finding appropriate land, influenced by factors such as soil quality, terrain, and pollution, necessitates sophisticated tools and specialized knowledge. This study utilizes SPARQL queries against the KnowWhereGraph (KWG) and data from the gSSURGO dataset, processed through ArcGIS, to streamline this task. Through the integration of these resources, this research seeks to simplify the access and interpretation of critical data dispersed across various entities, enabling the achievement of efficient land use. The outcomes of this research are anticipated to contribute to enhanced food security, economic growth, and increased access to renewable energy, aligning with local and global sustainability goals. Focusing initially on Ohio-where the funding university is located-the methodologies developed could be adapted for broader geographical applications, making this approach a scalable solution for future land use planning.

Keywords: KnowWhereGraph · Knowledge Graph · SPARQL · Agriculture · ArcGIS · Land Use · gSSURGO · Sustainability

1 Introduction

The United States Department of Agriculture (USDA), local farmers, and the United Nations have expressed an interest in maximizing efficient land use for future farm and renewable energy developments [1,6,13]. However, finding suitable land for these projects can be arduous and depends on many factors, such as soil quality, terrain, pollution, and other elements that influence crop growth or energy efficiency [1]. Additionally, the data that should be used for determining the ideal location for these future land use projects are dispersed among many different entities, require specialized tools to navigate through, or may be difficult to understand without a background in a specific field of study [14]. Optimizing the data aggregation and analysis process for users will alleviate the time-consuming and resource-intensive task of identifying the appropriate locations for future agriculture land use projects. This may result in several

S. Tiwari et al. (Eds.): AI4S 2024, CCIS 2243, pp. 16–26, 2025.
https://doi.org/10.1007/978-3-031-81369-6_2

additional long-term benefits such as enhanced food security, economic growth, and increased access to reusable energy, which are goals that align with both local and global communities in creating a more sustainable future [1,6,13].

This research is intended to provide an avenue for efficiently exploring and identifying potential high quality land for agriculture and renewable energy use. It will use SPARQL queries against the relevant aggregated data in KnowWhere-Graph (KWG) and information exported from the gSSURGO dataset manipulated in ArcGIS. Using SPARQL queries against the KWG database, the appropriate relationships between the available data can be more easily identified and used to generate a relevant dataset. This dataset will aid users in identifying the most appropriate locations for their future land development projects, more specifically in the realm of agriculture and energy land use. The users of this data can be lawmakers, local farmers, other government entities, or researchers.

This paper intends to answer whether knowledge graphs, specifically KWG, can be used to gather the necessary data for processing with other tools like ArcGIS Pro, to efficiently identify prime locations for future land use projects for lawmakers and local farmers. We hypothesize that KWG can provide the valuable data required for processing with geographic information system software, aiding in the strategic decision making process for future land use development. This research will focus on the relevant data for the state of Ohio, which is located in the Midwest Region of the USA, where land use and development is of high concern [1,6]. However, the schema for this knowledge graph is reusable, and the SPARQL queries can be easily modified to determine the best land use for other geographic locations, provided of course, that the data exists in KWG.

All the resources used in this research, such as SPARQL queries, code, and ArcGIS project files, can be found in our dedicated Kastle-Lab repository: https://github.com/kastle-lab/KWG-LandUse [11].

The arrangement of this paper is as outlined: Sect. 2 will include the related work that this project builds upon or utilizes. Section 3 will outline the tools and datasets utilized throughout the research and the results. Sections 4 and 5 will discuss the results, future work, limitations and the conclusion of this research.

2 Background and Related Work

2.1 KnowWhereGraph

Ontologies, knowledge graphs (KGs), and linked data are tools that can and have been utilized to answer innumerable questions in various scientific fields and domains [10,14]. Agriculture is one of the domains that is actively exploring the use of KGs to solve several issues, one issue being semantic interoperability, which is loosely defined as the ability to incorporate resources and data from various ontologies and perspectives (semantics) [7,10]. Fortunately, KWG does an excellent job of solving the semantic interoperability problem and allows users to utilize vast amounts of information from various sources, including those within the domain of Agriculture, which is in the interest of this study [8,14]. In an article published in AI Magazine on the topic of KWG, Janowicz, K. et al.

state that data retrieval and processing consume most resources in projects that require large amounts of data to make strategic decisions [8,10]. KWG contains over 30 integrated datasets and has been shown to alleviate this heavy resource consumption and allow users to retrieve the information needed to make the appropriate strategic decisions in a reasonable time frame [8,14]. Additionally, KWG already contains the necessary geospatial data required for this research and has customized tools specifically designed for usage in ArcGIS Pro [8,16]. The above reasons ultimately influenced the decision to use KWG, SPARQL queries, and ArcGIS Pro for this research.

2.2 Semantic Web For Urban Planning

A research team led by Heidi Silvennoinen from the Singapore-ETH Centre is currently exploring the efficacy and usefulness of semantic web technologies in urban development. Their approach consisted of creating a series of competency questions that could be answered through quantifiable results from a quarriable knowledge graph that they created, reusing existing ontologies when necessary. Near the end of their ontology creation process, the research team linked geospatial data from a variety of databases, a necessary step to effectively answer their competency questions. Using a Python script and a SPARQL wrapper with existing libraries to assist in visualizing the data, the team demonstrated the potential usefulness of their queried results [15]. The approach that the research team from the Singapore-ETH Centre took is similar to the one taken in our research and shows a potential avenue for success in efficient and sustainable land use development using a comparable methodology.

2.3 Literature Gap

Currently, the authors are unaware of, or there is little research literature on the topic of using KGs for identifying prime farmland for sustainable land use. The authors believe that it is important to address this issue as food security and sustainability, as they are primary concerns for world and local governments, as mentioned previously in Sect. 1. With this research, we intend to showcase how KGs (KWG) can be utilized as a technique for exploring solutions to the issues that exist within the domain of agriculture. Thus, making available more potentially viable pathways regarding sustainable land use in agriculture.

3 Approach

To further explore the use of semantic web technology for sustainable agriculture land use development we first created several competency questions that we wanted to answer. The example competency question used for this paper is: How much uncultivated prime farmland exists in Shelby and Madison County, Ohio? Then, instead of making our own ontology and knowledge graph, we used the existing graph inside of KWG, which is already queryable and removes the laborious and time-consuming work required in designing ontologies. In the same way as the researchers on Heidi Silvennoinen's team in Singapore did, we created SPARQL queries to gather useful data from KWG to help answer our competency questions [15]. How our approach differs from the methods used by the Singapore research team, is we applied a Python script inside the geographic information software (GIS) ArcGIS Pro, to process the data queried from KWG to create geometries that can be projected onto an existing map [15]. Then, using the existing tools inside the GIS software, that data was combined with other available maps from the ESRI Living Atlas database to reveal overlaps, making the data easy to visualize [4,5,12].

3.1 Extraction

Translating the above competency question from natural language into a SPARQL query requires consulting the KWG schema and performing preliminary queries to get the information necessary for the final query. To begin, it was necessary to identify the Uniform Resource Identifier (URI) for the counties of interest, which required querying KWG for the administrative region where the county data resides and identifying how the data is structured, as seen in Fig 2. Then it was necessary to reference the KWG schema and determine the relationship network between entities that would produce usable data for ArcGIS (Fig 1). For this case, the solution was to acquire Shelby and Madison County geometries that overlapped with the geometries of the gSSURGO data labeled as prime farmland. Refer to Fig 3 to see the SPARQL query for retrieving the polygons of Shelby County from KWG.

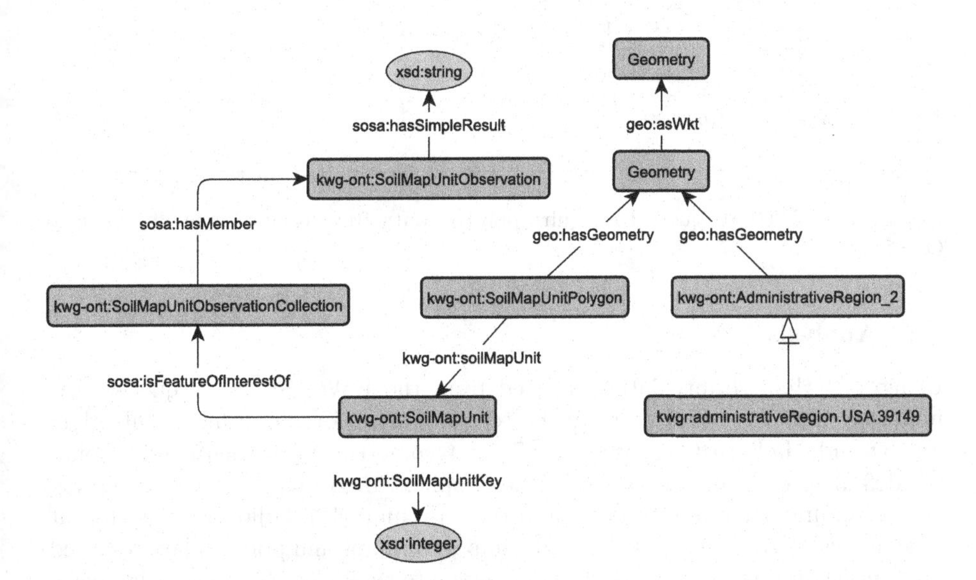

Fig. 1. The KnowWhereGraph relationships for obtaining the necessary geometries from the SPARQL queries [9].

```
PREFIX rdfs: <http://www.w3.org/2000/01/rdf-schema#>
PREFIX kwg-ont: <http://stko-kwg.geog.ucsb.edu/lod/ontology/>

SELECT ?reg ?rdf ?label where
{
    ?reg ?rdf kwg-ont:AdministrativeRegion_2 ;
        rdfs:label ?label .
     FILTER (?label="Shelby County, Ohio")
} LIMIT 10
```

Fig. 2. The SPARQL query to identify the URI for Shelby County Ohio

```
PREFIX sosa: <http://www.w3.org/ns/sosa/>
PREFIX rdf: <http://www.w3.org/1999/02/22-rdf-syntax-ns#>f
PREFIX kwg-ont: <http://stko-kwg.geog.ucsb.edu/lod/ontology/>
PREFIX rdfs: <http://www.w3.org/2000/01/rdf-schema#>
PREFIX geo: <http://www.opengis.net/ont/geosparql#>
PREFIX kwgr: <http://stko-kwg.geog.ucsb.edu/lod/resource/>

SELECT DISTINCT ?wkt  WHERE
{
    BIND(kwgr:administrativeRegion.USA.39149 AS ?shelbyCounty)
    ?shelbyCounty rdfs:label ?county .
    ?shelbyCounty geo:hasGeometry ?shelbyGeo .
    ?smuo a kwg-ont:SoilMapUnitObservation ;
        sosa:hasSimpleResult ?result .
    FILTER(?result = "All areas are prime farmland")
    ?smuoc a kwg-ont:SoilMapUnitObservationCollection ;
        sosa:hasMember ?smuo .
    ?smu a kwg-ont:SoilMapUnit ;
        kwg-ont:soilMapUnitKey ?smukey ;
        sosa:isFeatureOfInterestOf ?smuoc .
    ?smup a kwg-ont:SoilMapUnitPolygon ;
        kwg-ont:soilMapUnit ?smu ;
        geo:hasGeometry ?geoObj .
    ?shelbyGeo geo:sfOverlaps ?geoObj .
        ?geoObj geo:asWKT ?wkt .
 }
```

Fig. 3. The SPARQL query to acquire polygon data from KWG for Shelby County Ohio.

3.2 Analysis

Using a Python script, data extracted from the KWG SPARQL queries was imported into ArcGIS Pro to generate geometries for all prime farmland in Shelby and Madison County, Ohio. These Geometries were then overlaid onto the default project map [4,5]. Next, the Cropland USA map from ESRI's Living Atlas was imported and overlaid onto the base map [12]. Following this step, it was important to verify that any of the imported maps and polygon layers shared the same spatial referencing coordinate system to avoid coordinate misalignment. Initially, the coordinate systems did not match and had to be converted to the Albers Conical Equal Area system. Except for the default World Topological and

Hillshade maps as they are simply used as a base map and were not intended to provide meaningful insights into the data [4,5]. Afterward, the cropland data was clipped out using the imported geometries with the ArcGIS Geoprocessing Raster Clipping Tool. This transformation created a geometric map highlighting both the existing and absent cropland within the designated prime farmland areas for Shelby County (Fig 4) and Madison County (Fig 6).

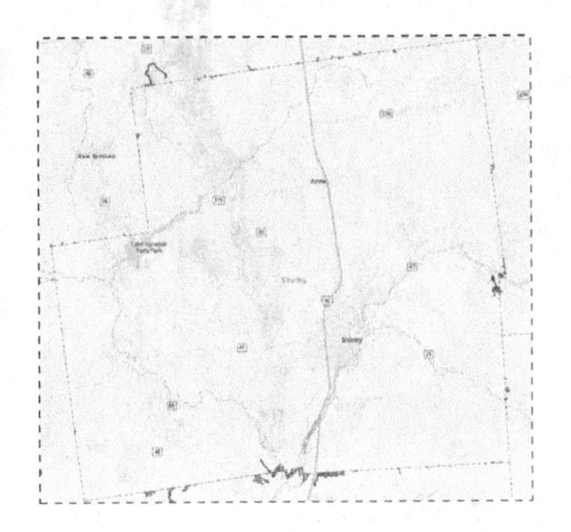

Fig. 4. This ArcGIS map section displays the geometries of prime farmland in Shelby County Ohio, overlapped with cropland data. Black areas indicate uncultivated prime farmland. Refer to Fig 5 for a closer inspection of the data.

Fig. 5. This is a zoomed-in view of the bottom center section of the results from Fig 4

4 Results

Once the map transformation was completed valuable data was extracted from the attribute table generated from the preceding processes (see Tables 1 and 2). The data includes the type of crops in the designated prime farmland area and the number of cells occupying the geometries. Each cell in this rasterized layer represents an area of $900\,\mathrm{m}^2$. This data facilitated the calculation of the total area of uncultivated prime farmland by multiplying the number of cells that occupy the uncultivated section of the raster layer by the area of a cell. Then the results were converted from square meters to acres by dividing by approximately 4,047 m^2.

$$900\,\mathrm{m}^2 \times 1923 = 1.7307 \times 10^6\,\mathrm{m}^2 \tag{1}$$

$$\frac{1.7307 \times 10^6\,\mathrm{m}^2}{4.04685642 \times 10^3\,\mathrm{m}^2} \approx 427.81\,\mathrm{acres} \tag{2}$$

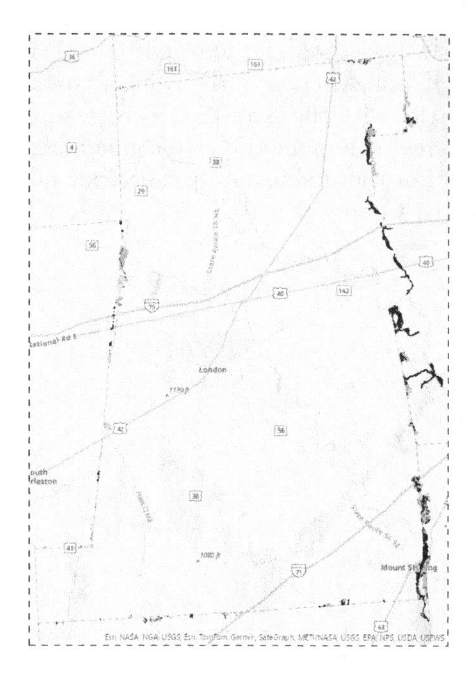

Fig. 6. This ArcGIS map section displays the geometries of prime farmland in Madison County Ohio, overlapped with cropland data. Black areas indicate uncultivated prime farmland. Refer to Fig 7 for a closer inspection of the data.

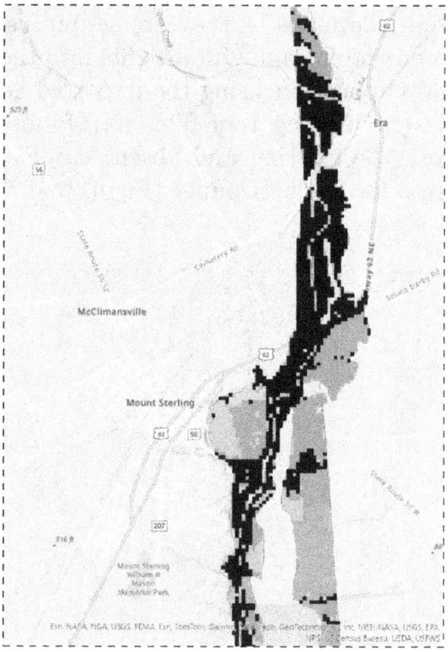

Fig. 7. This is a zoomed-in view of the bottom right corner of the results from Fig 6

ArcGIS Pro streamlines this calculation by allowing the user to add a numeric field to the attribute table and perform the formula operations inside that field. With the information extracted from ArcGIS, we calculated the total percentage of prime farmland in each county of our research by taking the acreage of prime farmland and dividing it by their respective county's total acreage. According to the United States Census Bureau (USCB) the total area of Shelby County, Ohio is 260,928 acres (173 hectares) and Madison County is 298,688 acres (712 hectares) [3]. In both cases, the total percentage of prime farmland constitutes less than one percent of the total area of each county.

Table 1. This attribute table displays various useful information of the crop data inside the geometries of Madison County, Ohio.

OBJECTID *	Count	Class_Name	Popup Text	Area
1	7912	Uncultivated	no cultivated crops	1759.588
6	4407	Soybeans	soybeans	980.0941
2	2349	Corn	corn	522.4055
16	840	Winter Wheat	winter wheat	186.8117
28	119	Alfalfa	alfalfa	26.46499
29	108	Other Hay/Non Alfalfa	other hay (but not alfalfa)	24.01864
18	103	Dbl Crop WinWht/Soybeans	a double crop of winter wheat and soybeans	22.90667
52	5	Fallow/Idle Cropland	fallow or idle cropland	1.111974
50	4	Sod/Grass Seed	sod or grass seed	0.889579
41	2	Cucumbers	cucumbers	0.44479
49	2	Clover/Wildflowers	clover and/or wildflowers	0.44479
20	1	Oats	oats	0.222395

After performing extraction and analysis it is evident that combining KWG with other tools that deal with geographic information systems, such as ArcGIS Pro, facilitates the identification of future land use development. Due to the restrictions on the capability of KWG, such as the five minute query timeout limit, we were only able to successfully query so many counties. However, The useful information that results from this study is a map with interpretable geometries that contain valuable data for calculations used to determine the total area of uncultivated prime farmland in Shelby and Madison County, Ohio. In this study, it was revealed that the total area of uncultivated prime farmland in said counties approximates to slightly over 427 acres for Shelby County and as much as 1,759 acres in Madison County.

Table 2. This attribute table displays various useful information of the crop data inside the geometries of Shelby County, Ohio.

OBJECTID *	Count	Class_Name	Popup Text	Acres
1	1923	Uncultivated	no cultivated crops	427.6653
6	1283	Soybeans	soybeans	285.3326
2	784	Corn	corn	174.3576
16	172	Winter Wheat	winter wheat	38.25191
28	111	Alfalfa	alfalfa	24.68583
29	23	Other Hay/Non Alfalfa	other hay (but not alfalfa)	5.115081
18	16	Dbl Crop WinWht/Soybeans	a double crop of winter wheat and soybeans	3.558317
11	1	Pop or Ornamental Corn	popcorn or ornamental corn	0.222395

It should be noted that the topographic map which displays the county boundaries was not aligned to the exact projection coordinate system as the other layers. Therefore, it was not used to make any determinations in this study. Due to the nature of projected rasterized imagery data interacting with imported geometries, it should be mentioned that these calculations are intended for approximations rather than precise evaluation.

4.1 Extrapolation

According to the Economic Research Service at the USDA, it is estimated that each acre in Ohio that is cultivated with corn may yield up to twenty-seven U.S. Dollars (USD), after all expenses are accounted for [2]. In the two counties that our research focused on there is a cumulative total of 2,187 acres of uncultivated prime farmland. Multiplying the cumulative total of prime farmland with the value produced per acre of corn, results in an estimated value of $59,049. Ohio has a total of eighty-eight counties, each assumed to have its areas of accessible uncultivated prime farmland that hold significant potential value in terms of USD. We assume that there are hundreds of thousands of dollars worth of uncultivated farmland that exists in the state of Ohio. However, further research and examination of data are needed to substantiate this assumption.

5 Conclusion

In this study, we aimed to prove that KnowWhereGraph (KWG) can be integrated with other geographic information system (GIS) processing tools to assist users in efficiently identifying key locations for future land use development. Knowledge graphs are excellent mechanisms for integrating heterogeneous data from various sources. This is very necessary in the case of land use and development, and especially the assessment thereof, due to the fact that it requires an immense amount of contextual knowledge. Appropriately, KnowWhereGraph, in particular, is a convenient source of knowledge, due to its focus on geospatial related data.

In this paper, we have demonstrated that we can gather the necessary data from KWG for processing with other tools like ArcGIS Pro to efficiently identify key locations for future land use development. Using GIS software allowed us to make valuable insights from the imported KWG data, such as the approximate land coverage of uncultivated prime farmland in Shelby and Madison County, Ohio. Effective utilization of this data required the use of Python scripting to make the necessary conversion required for the data to be accurate with other imported data used for calculations. The implications of these findings indicate that KWG is indeed a valuable resource that be used in making strategic decisions about land use development. Though the scope of this research only focused on two counties within the state of Ohio, USA, the methodology that was used can be scaled to the state or country level, though this would require significantly more time and resources to accomplish.

Limitations and Future Work

We have identified some next steps to grow our work from these preliminary stages.

- We wish to demonstrate a wider applicability of the approach (i.e., using KWG as both a convenient source of data, as well as for ad hoc geoinference) by choosing additional regions of interest (e.g., other administrative or development regions).
- KnowWhereGraph is frequently updating its graph. Currently, we expect that additional data (and some schema changes) will occur by the end of the calendar year, which will allow us to explore additional layers to incorporate into our analysis.
- Eventually we seek to test if results like those discovered in this study could be identified through the use of natural language by non-expert users through a custom trained Large Language Model (LLM).
- Currently, the SPARQL queries will sometimes result in a Gateway Time-Out when trying to procure the data from KWG's servers. We suspect that these queries can be altered to reduce complexity, allowing for more efficient server-side computations and resulting in shorter data retrieval times.

Acknowledgments. This work was, in part, funded by the National Science Foundation (NSF) under Grant #2333532; Proto-OKN Theme 3: An Education Gateway for the Proto-OKN. Any opinions, findings, and conclusions or recommendations expressed in this material are those of the authors and do not necessarily reflect the views of the NSF. We would like to thank Antrea Christou, Alexis Ellis, and the rest of the Kastle Lab at Wright State University for their valuable feedback and suggestions toward improving the quality of this paper.

Disclosure of Interests. The authors affirm that no conflicts of interest, either financial or personal, could have influenced the reported findings.

References

1. of Agriculture, U.S.D.: USDA strategic plan fiscal years 2022-2026 (2022). https://www.usda.gov/sites/default/files/documents/usda-fy-2022-2026-strategic-plan.pdf
2. of Agriculture Economic Research Service, U.D.: USDA ERS - commodity costs and returns (2024). https://www.ers.usda.gov/data-products/commodity-costs-and-returns
3. Bureau, U.S.C.: explore census data (2020)., https://data.census.gov/
4. Environmental Systems Research Institute, I.: World hillshade (2015). https://www.arcgis.com/home/item.html?id=1b243539f4514b6ba35e7d995890db1d#
5. Environmental Systems Research Institute, I.: World topographic map (2017). https://www.arcgis.com/home/item.html?id=7dc6cea0b1764a1f9af2e679f642f0f5
6. Food, of the United Nations, A.O.: Sustainable land management — land & water — food and agriculture organization of the united nations — land & water — food and agriculture organization of the united nations (2023). https://www.fao.org/land-water/land/sustainable-land-management/en/
7. Heflin, J., Hendler, J.: Semantic interoperability on the web. In: Proceedings of Extreme Markup Languages, vol. 2000, pp. 111–120 (2000). https://www.researchgate.net/profile/Jeff-Heflin/publication/242233182_Semantic_Interoperability_on_the_Web/links/0a85e53b54ffb40fcc000000/Semantic-Interoperability-on-the-Web.pdf

8. Janowicz, K., et al.: Know, know where, KnowWhereGraph: a densely connected, cross-domain knowledge graph and geo-enrichment service stack for applications in environmental intelligence. AI Mag. **43**(1), 30–39 (2022). https://doi.org/10.1609/aimag.v43i1.19120
9. Knowwheregraph. . https://knowheregraph.org/
10. López-Morales, J.A., Martínez, J.A., Skarmeta, A.F.: Digital transformation of agriculture through the use of an interoperable platform. Sensors **20**(4), 1153 (2020). https://doi.org/10.3390/s20041153
11. McCain, M., Kandula, R., Shimizu, C.: KnowWhereGraph for land use optimization: achieving sustainability and efficiency. https://github.com/kastle-lab/KWG-LandUse
12. (NASS), N.A.S.S.: USDA NASS cropland data layer (2008). https://landscape11.arcgis.com/arcgis/rest/services/USA_Cropland/ImageServer
13. Nations, U.: Transforming our world: the 2030 agenda for sustainable development (2015). https://sdgs.un.org/2030agenda
14. Shimizu, C., et al.: The KnowWhereGraph ontology: a showcase (2023). https://ceur-ws.org/Vol-3637/paper46.pdf
15. Silvennoinen, H., et al.: A semantic web approach to land use regulations in urban planning: the Ontozoning ontology of zones, land uses and programmes for Singapore. J. Urban Manage. **12**, 151–167 (2023). https://doi.org/10.1016/j.jum.2023.02.002, https://www.sciencedirect.com/science/article/pii/S2226585623000067
16. Zhu, R.: Geospatial knowledge graphs (2024). https://doi.org/10.48550/arXiv.2405.07664

A Multiple Criteria-Based Customer Segmentation and Recommender System for Rural Farmers

Patience U. Usip[1,2](✉), Ubong Etop Gibson[1], and Samuel S. Udoh[3]

[1] Department of Computer Science, University of Uyo, Uyo, Nigeria
patienceusip@uniuyo.edu.ng
[2] TETFund Center of Excellence in Computational Intelligence Research, University of Uyo, Uyo, Nigeria
[3] Department of Data Science, University of Uyo, Uyo, Nigeria

Abstract. Agriculture plays an important role in any nation's economy. Agricultural marketing brings producers (farmers often in rural areas) and customers (customers often in rural areas) together at various location/distribution centres. Farm produce will be distributed to the location/centres based on the customers' needs and hence requires that customers are segmented based on the needs and nearness to distribution centres. Segmenting customers and recommending other farm produce available at a distribution centre is based on multiple criteria including user demographics, and historical behaviour. This paper is aimed at segmenting the available customers and recommending available farm produce at various distribution centres for easy access for or by the rural farmers. The K-mean Algorithm is used for customer segmentation, while the association rule apriori algorithm for farm produce recommendation. The dataset of 300 farmers in different villages in Oruk Anam Local Government Area, Akwa Ibom State were collected, followed by the model implementation using python programming language in jupyter notebook and other web technologies. The Apriori algorithm was used to find the frequent item sets based on clusters from the RFM-K Means model. The frequent itemsets for different clusters (0,1,2) were derived and minimum support of 0.03 for cluster 2, 0.02 for cluster 1 and 0.01 for cluster 0 were applied to find the frequent itemsets (The algorithm grouped buyers and sellers into three clusters). Minimum threshold of 1, minimum confidence score of 0.7 and minimum lift score of =>4, 2, 2 for cluster 2, 1, 0 respectively were applied to the association rules to achieve a better result. The minimum support threshold determines the frequency required for a cluster to be considered relevant. Confidence in this represents the reliability of the association rules. A higher confidence threshold means more reliable recommendations. This work provides the platform for accessibility to rural farmers and the farm produce.

Keywords: Knowledge mining · customer segmentation · product recommender system · e-agriculture

S. Tiwari et al. (Eds.): AI4S 2024, CCIS 2243, pp. 27–42, 2025.
https://doi.org/10.1007/978-3-031-81369-6_3

1 Introduction

The agriculture sector is a crucial pillar of the global economy, encompassing a wide demographic and exerting a significant influence on the socio-economic fabric of numerous countries worldwide [1]. At the heart of this sector, agricultural marketing serves as a vital link between producers (farmers) and consumers (customers), focusing on both the final agricultural produce and the supply of necessary inputs to farmers [2]. Several researches have been recently carried out by authors to boost agricultural marketing such as development of a machine learning chatbot [3] and its usability evaluation [4]. The evolution of e-commerce within the realm of e-agriculture has introduced transformative changes, offering considerable benefits to customers and businesses on a global scale. E-commerce facilitates online transactions, services, and information exchange, fostering a rapid and efficient digital marketplace [5].

In the contemporary competitive business landscape, companies are driven to gain a competitive edge by understanding and catering to the diverse needs and preferences of their customers. Personalized marketing and customer experiences have become indispensable strategies for enhancing customer satisfaction, loyalty, and revenue growth. To achieve this, effective customer segmentation has emerged as a powerful tool, allowing companies to tailor their marketing strategies to specific customer groups with similar attributes or behaviors.

Traditional approaches of treating all customers uniformly are no longer effective in a world where personalised interactions are expected. Customer segmentation enables businesses to gain insights into the unique needs, preferences, and motivations of distinct customer groups. This understanding facilitates the design of targeted marketing strategies and the delivery of personalised experiences.

Recommender systems play a pivotal role in providing personalised experiences by leveraging algorithms and data analysis techniques to suggest relevant items or content based on users' past interactions and preferences. These systems analyse historical data and patterns to generate personalised recommendations, thereby enhancing user engagement and satisfaction.

Moreover, due to the challenges posed by an abundance of product options, customers may struggle to identify suitable products quickly when shopping online. Product recommendation systems address this issue by logically displaying only relevant products, improving the shopping experience by reducing time spent using content-based, collaborative filter-based, or hybrid as the main types of recommendation systems [6].

The integration of artificial intelligence (AI) and machine learning (ML) algorithms has brought about a digital revolution in agriculture, transforming managerial functions into intelligent systems. Machine learning, a branch of AI, holds great promise for addressing challenges in developing knowledge-based agricultural systems [7].

Data mining and clustering techniques, such as classification, association rules, and hierarchical clustering algorithms, play a significant role in customer segmentation and product recommendation systems [8]. This paper aims to leverage machine learning approaches to enhance customer segmentation and recommender systems, designing a framework that integrates these components to provide highly targeted and personalised recommendations to customers and farmers in rural areas. The goal is to empower farmers with tailored marketing strategies, thereby enhancing customer engagement, satisfaction, loyalty, and overall business success.

Agriculture holds a crucial role in the global economy, impacting the socio-economic fabric of many countries. The emergence of e-commerce, particularly in the realm of e-agriculture, has brought significant benefits through online transactions, services, and information exchange. In the competitive business landscape, personalised marketing and customer experiences are essential for satisfaction, loyalty, and revenue growth. Customer segmentation, dividing customers based on characteristics or behaviours, is a powerful strategy, enhanced by recommender systems that suggest tailored content or products based on user preferences. The increasing challenge of product options demands effective product recommendation systems, utilising bidirectional communication for optimised and personalised suggestions. With the integration of AI and ML, agriculture has undergone a digital revolution, particularly in customer segmentation, a practice of dividing customers into distinct groups based on specific characteristics or behaviours and product recommendation systems. This paper aims to leverage machine learning for customer segmentation and recommender systems, creating a framework for highly targeted and personalised recommendations to customers and farmers in rural areas, thereby enhancing marketing strategies and overall business success.

2 Related Literature

Udokwu et al. [6] proposed a system that applied a hybrid system of machine learning association and clustering algorithms to implement a product recommendation system that showed associations that existed in products and unique customer profiles linked to these associations. The machine learning methods such as k-means clustering and Apriori association rule were evaluated with a case of a hygiene product retailer in Austria.

Kamble and Shaha [9] developed a system that was focused on improving the performance of recommender systems by using data mining techniques. The machine learning algorithm such as Support Vector Machine was used for the product recommendation system. They conducted an experiment to evaluate their model on each type of product data set such as Grocery, Beauty, Cold drinks, watches, lockers, Gourmet Foods etc. and compared them with the traditional algorithm. The results showed the model was an improvement in making recommendations.

Zhao and Keikhosrokiani [10] proposed a novel data science life cycle and process model. Their model used Recency, Frequency, and Monetary (RFM) analysis methods with the combination of various analytics algorithms which were utilised for sales prediction and product recommendation through user behaviour analytics. RFM analysis method was utilised for segmenting customer levels in the company to identify the importance of each level. For the purchase prediction model, XGBoost and Random Forest machine learning algorithms were applied to build prediction models and 5-fold Cross-Validation method was utilised to evaluate them for the product recommendation model, the association rules theory and Apriori algorithm were used to complete basket analysis and recommend products according to the outcomes. The result of the XGBoost model achieved better performance and better accuracy with an F1-score around 0.789. The recommendation model provided good recommendation results and sales combinations for improving sales and market responsiveness.

Shi et al. [8] proposed a system that was called an intelligent recommendation methodology, and they were able to get further effectiveness and quality of recommendations when applied to an Internet shopping mall. Recency, Frequency, Monetary model with Fuzzy Clustering Method were applied to the loyalty of customers (or customer lifetime value) to rank them.

3 Data Collection, Preprocessing and Clustering

The data from 300 training samples was collated from farmers in different villages in Oruk Anam local government area and was used for data analysis (customer segmentation and produce recommendation). The well-structured questionnaire with sections of feedback data, was administered to farmers and their feedback was collated on the farm produce sales. The data of these variables were given as input. Some features or attributes based on the questions were selected and extracted from the feedback and the dataset was pre-processed using Microsoft Excel Application in order to remove impurities. The dataset in csv (comma separated values) format was uploaded to Jupyter Lab for analysis and model building with RFM analysis, K-means and Apriori algorithms.

The dataset was cleansed by removing incomplete or duplicate instances as given in Table 1. In this step only those features needed were selected which were required for analysis.

Table 1. Dataset of buyers' feedback (CSV file)

	RESPID	AGE	GENDER	MSTATUS	VILLAGE	PRODUCE	TYPE	LASTPURCHASE	QUANTITY	PRICE	REVENUE
1	F002	0	M	M	IKOT INUEN	PALM OIL	W	2023-10-25	10	20000	200000
2	F001	0	M	M	IKOT INUEN	PALM OIL	R	2023-10-25	9	500	4500
3	F002	0	M	M	IKOT INUEN	PALM OIL	R	2023-10-25	10	500	5000
4	F001	0	M	M	IKOT INUEN	PALM OIL	R	2023-10-25	5	500	2500
5	F005	0	M	M	IKOT INUEN	PALM OIL	W	2023-10-25	8	20000	160000
6	F002	0	M	M	IKOT INUEN	PALM KERNEL	W	2023-10-25	8	10000	80000
7	F006	0	M	M	IKOT INUEN	CASSAVA	R	2023-10-25	9	100	900
8	F003	0	M	M	IKOT INUEN	WHEAT	R	2023-10-25	5	100	500
9	F007	0	M	M	IKOT INUEN	WHEAT	R	2023-10-25	7	100	700
10	F003	0	M	M	IKOT INUEN	COWPEA	R	2023-10-25	8	250	2000
11	F001	0	M	M	IKOT INUEN	COWPEA	W	2023-10-25	8	24000	192000
12	F008	0	M	M	IKOT INUEN	BEANS	W	2023-10-25	6	30000	180000
13	F005	0	M	M	IKOT INUEN	BEANS	R	2023-10-25	5	250	1250
14	F004	0	M	M	IKOT INUEN	BEANS	R	2023-10-25	9	250	2250
15	F003	0	M	M	IKOT INUEN	BEANS	R	2023-10-25	9	250	2250
16	F006	0	M	M	IKOT INUEN	BEANS	W	2023-10-25	6	30000	180000
17	F008	0	M	M	IKOT INUEN	MAIZE	W	2023-10-25	10	5000	50000
18	F005	0	M	M	IKOT INUEN	PLANTAIN	W	2023-10-25	8	3000	24000
19	F004	0	M	M	IKOT INUEN	PLANTAIN	R	2023-10-25	5	500	2500
20	F007	0	M	M	IKOT INUEN	PLANTAIN	R	2023-10-25	10	500	5000
21	F004	0	M	M	IKOT INUEN	PLANTAIN	R	2023-10-25	5	500	2500

nple 0 1 csdat

After the pre-processing, Table 2 shows the result of the dataset.

Table 2. Dataset details

RangeIndex: 300 entries, 0 to 299
Data columns (total 11 columns):

#	Column	Non-Null Count	Dtype
0	RESPID	300 non-null	object
1	AGE	300 non-null	object
2	GENDER	300 non-null	object
3	MSTATUS	300 non-null	object
4	VILLAGE	300 non-null	object
5	PRODUCE	300 non-null	object
6	TYPE	300 non-null	object
7	LASTPURCHASE	300 non-null	object
8	QUANTITY	300 non-null	int64
9	PRICE	300 non-null	int64
10	REVENUE	300 non-null	int64

dtypes: int64(3), object(8)
memory usage: 25.9+ KB

Demography of Respondents (Farmers/Buyers)

Figures 1, 2, 3 and 4 give the charts of distribution of attributes, which were extracted based on the questions that were given to the farmers/buyers to give their responses.

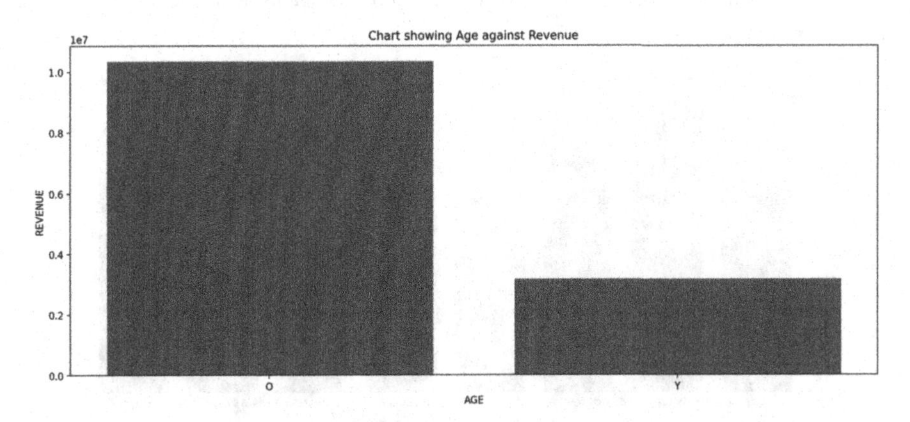

Fig. 1. Chart of distribution of Age against Revenue

Figure 1 shows that farmers/buyers of the old age group purchase more farm produce than the young age group.

Figure 2 shows that farmers/buyers of female gender purchase more farm produce than the male gender.

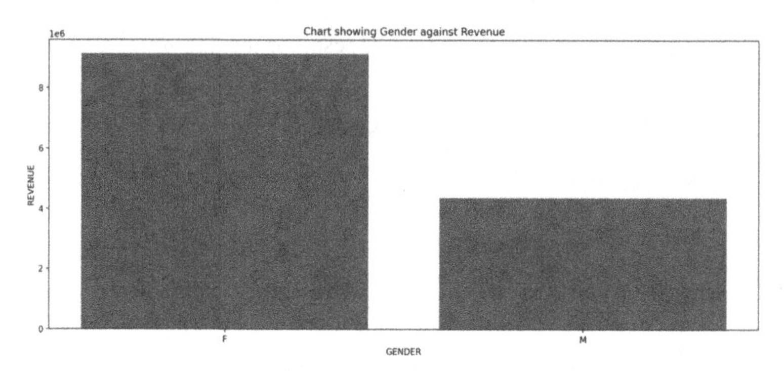

Fig. 2. Chart of distribution of Gender against Revenue

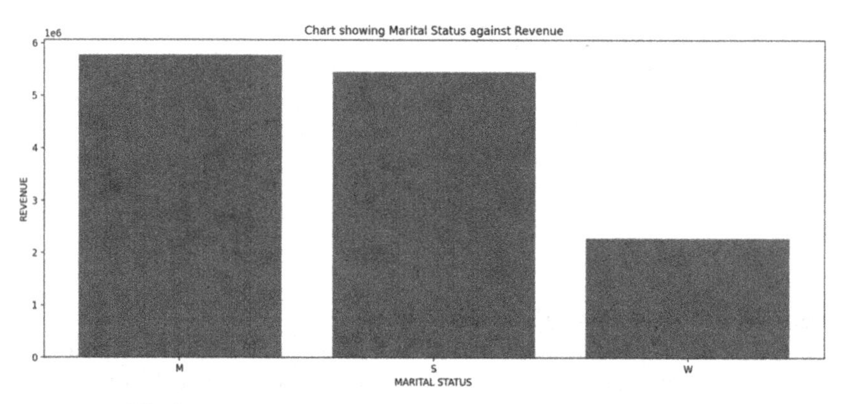

Fig. 3. Chart of distribution of Marital Status against Revenue

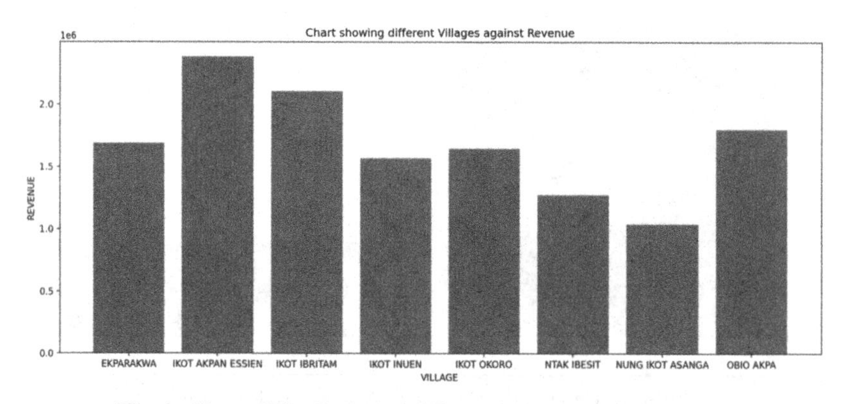

Fig. 4. Chart of distribution of different Villages against Revenue

Figure 3 shows that farmers/buyers that are married purchase more than single and both married and single buy more compared to that of widows.

Figure 4 shows that farmers/buyers from Ikot Akpan Essien purchase more farm produce than any other villages. Other villages such as Ikot Ibritam, Obio Akpa and Ekparakwa, Ikot Okoro, Ikot Inuen, Ntak Ibesit and Nung Ikot Asanga purchase farm produce accordingly.

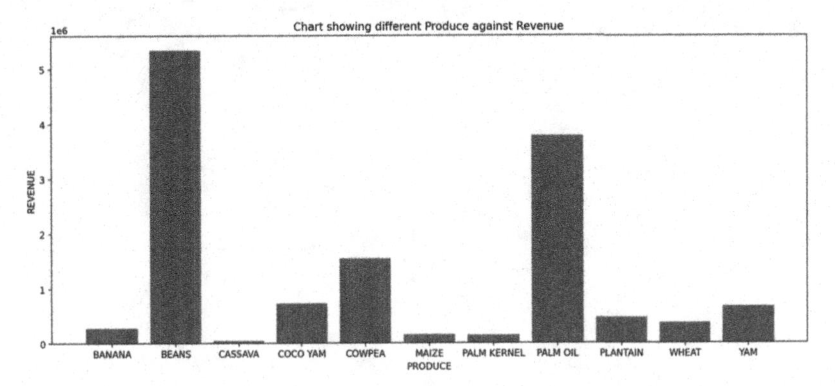

Fig. 5. Chart of distribution of different Produce against Revenue

Figure 5 shows that farmers/buyers purchase more beans than any other producer. They also purchase pail oil and cowpea and cocoyam more. Cassava and maize are the least purchased produce.

Recency, Frequency and Monetary Statistics

RFM analysis was carried out on the dataset and the Table 3 shows statistical values.

Table 3. RFM Statistics

S/N	Measurement parameters	R Values	F Values	M Values
1	Count	104.0	104.0	104.0
2	Mean	0.73	2.89	129892.79
3	Std	0.75	1.08	121690.81
4	Min	0.0	1.0	900.0
5	Max	2.0	7.0	535000

Table 3 shows the statistics of Recency, Frequency and Monetary analysis. It uses some measurement parameters such as count, mean, std, min and max.

RFM Plot

Figure 6 shows the distribution of recency, frequency and monetary against density. For recency, it reveals that recency increases at point 0 and decreases at point 2.

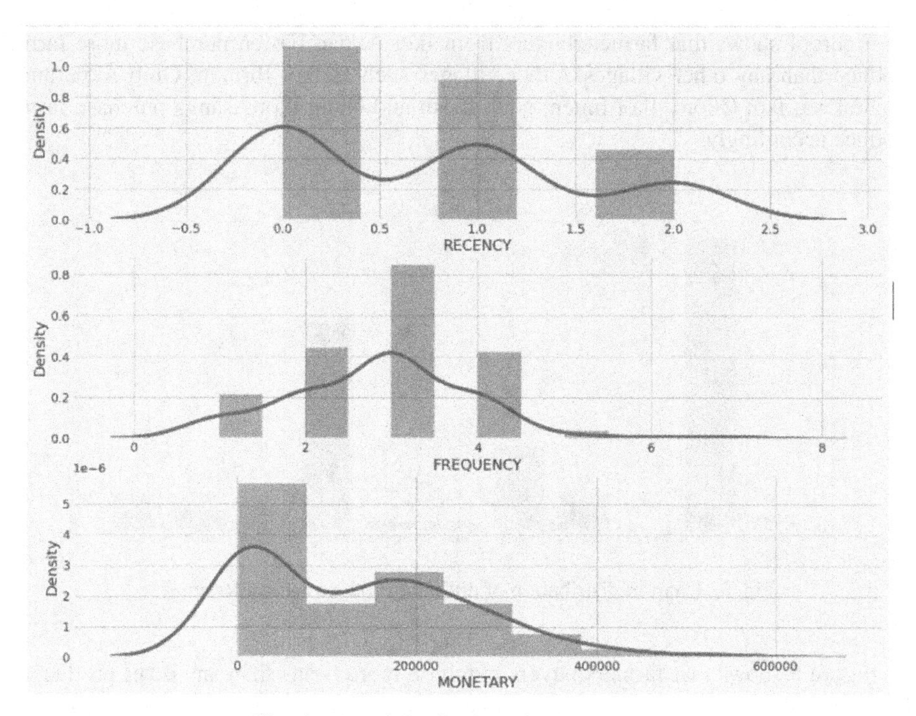

Fig. 6. Plot of distribution of RFM Analysis

For frequency, it reveals that frequency increases at point 2, 3 and 4, while decreases at point 0,1 and 5, 6, 7, and 8. While monetary reveals that purchases occur more within 0 and 200,000 than 200,000 and 600,000.

General Segment Result

After applying the RFM analysis to the dataset, RFM score was derived, which aided in the segmentation of instances into three (3) categories, such as High-rate buyers, Average-rate buyers and Low. Table 4 shows the general segmentation of customers.

Table 4. General Segment Result

	RESPID	LASTPURCHASE_DATE	RECENCY	FREQUENCY	MONETARY	R	F	M	RFM_Segment	RFM_Score	General_Segment
0	F001	2023-10-25	2	3	199000	1	2	3	123	6	Average-rate Buyers
1	F002	2023-10-25	2	3	285000	1	2	4	124	7	High-rate Buyers
2	F003	2023-10-25	2	3	4750	1	2	1	121	4	Average-rate Buyers
3	F004	2023-10-25	2	4	8450	1	4	1	141	6	Average-rate Buyers
4	F005	2023-10-25	2	4	260250	1	4	4	144	9	High-rate Buyers
...	...	...	...	...	...	...	...	...	...	...	...
99	F107	2023-10-26	1	2	4500	2	1	1	211	4	Average-rate Buyers
100	F108	2023-10-26	1	2	2100	2	1	1	211	4	Average-rate Buyers
101	F109	2023-10-26	1	1	24000	2	1	2	212	5	Average-rate Buyers
102	F110	2023-10-26	1	1	4500	2	1	1	211	4	Average-rate Buyers
103	F111	2023-10-26	1	2	273000	2	1	4	214	7	High-rate Buyers

104 rows × 11 columns

RFM Model with Cluster Result

To estrange instances into clusters, K-means algorithm was applied, with 3 as the number of clusters (k). The number of clusters was derived from the silhouette score, which shows that 3 was accepted as its value was close to 1. Figure 7 shows the Elbow curve.

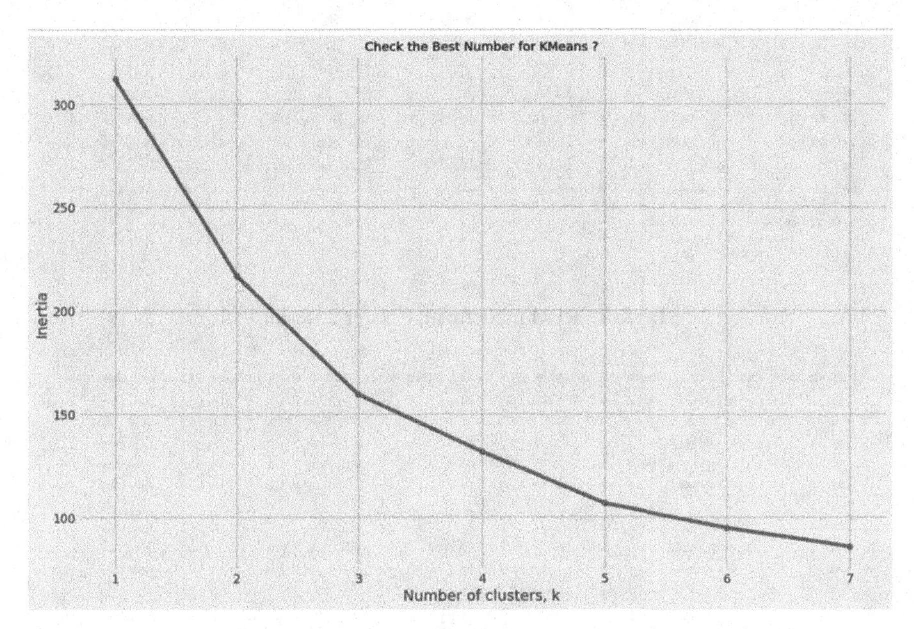

Fig. 7. Elbow Curve for number of cluster (k)

Table 5. RFM model with Cluster 0 Result

	RESPID	AGE	GENDER	MSTATUS	VILLAGE	PRODUCE	TYPE	LASTPURCHASE	QUANTITY	PRICE	...	RECENCY	FREQUENCY	MONETARY	R	F	M	RFM_Segment	RFM_Score	General_Segment	Cluster
111	F034	O	F	S	OBIO AKPA	PALM OIL	W	2023-10-26	6	20000	...	0	4	127150	2	4	3	243	9	High-rate Buyers	0
112	F034	O	F	S	OBIO AKPA	PALM OIL	R	2023-10-26	8	500	...	0	4	127150	2	4	3	243	9	High-rate Buyers	0
113	F034	O	F	S	OBIO AKPA	BEANS	R	2023-10-27	7	250	...	0	4	127150	2	4	3	243	9	High-rate Buyers	0
114	F034	O	F	S	OBIO AKPA	BANANA	R	2023-10-27	7	200	...	0	4	127150	2	4	3	243	9	High-rate Buyers	0
115	F036	O	F	S	OBIO AKPA	PALM OIL	R	2023-10-26	10	500	...	0	3	155900	2	2	3	223	7	High-rate Buyers	0
...	...	...	...	...	...	...	...	...	...	...	...	...	...	...	...	...	...	...	...	...	...
242	F083	Y	M	S	IKOT AKPAN ESSIEN	COCO YAM	R	2023-10-27	9	500	...	0	3	189500	2	2	3	223	7	High-rate Buyers	0
243	F081	Y	M	S	IKOT AKPAN ESSIEN	PALM KERNEL	R	2023-10-27	9	200	...	0	2	301800	2	1	4	214	7	High-rate Buyers	0
244	F081	Y	M	S	IKOT AKPAN ESSIEN	BEANS	W	2023-10-27	10	30000	...	0	2	301800	2	1	4	214	7	High-rate Buyers	0
245	F082	Y	M	S	IKOT AKPAN ESSIEN	COWPEA	R	2023-10-27	8	250	...	0	2	212000	2	1	4	214	7	High-rate Buyers	0
246	F082	Y	M	S	IKOT AKPAN ESSIEN	BEANS	W	2023-10-27	7	30000	...	0	2	212000	2	1	4	214	7	High-rate Buyers	0

110 rows × 22 columns

Table 5 shows the cluster result of zero (0) after applying the K-means algorithm.
Table 6 shows the cluster result of one (1) after applying the K-means algorithm.
Table 7 shows the cluster result of two (2) after applying the K-means algorithm.
Figure 8 shows the Recency, Frequency and Monetary plot in a 3D view.

The horizontal (x-axis) represent Frequency, the vertical (y-axis) represents the Monetary values, while the applical z-axis represent decency.

Table 6. RFM model with Cluster 1 Result

	RESPID	AGE	GENDER	MSTATUS	VILLAGE	PRODUCE	TYPE	LASTPURCHASE	QUANTITY	PRICE	...	RECENCY	FREQUENCY	MONETARY	R	F	M	RFM_Segment	RFM_Score	General_Segment	Cluster
3	F001	O	M	M	IKOT INUEN	PALM OIL	R	2023-10-25	9	500	...	2	3	199000	1	2	3	123	6	Average-rate Buyers	1
4	F001	O	M	M	IKOT INUEN	PALM OIL	R	2023-10-25	5	500	...	2	3	199000	1	2	3	123	6	Average-rate Buyers	1
5	F001	O	M	M	IKOT INUEN	COWPEA	W	2023-10-25	8	24000	...	2	3	199000	1	2	3	123	6	Average-rate Buyers	1
10	F006	O	M	M	IKOT INUEN	CASSAVA	R	2023-10-25	9	100	...	2	3	182500	1	2	3	123	6	Average-rate Buyers	1
11	F006	O	M	M	IKOT INUEN	BEANS	W	2023-10-25	6	30000	...	2	3	182500	1	2	3	123	6	Average-rate Buyers	1
...	...	...	...	...	...	...	...	...	...	...	...	...	...	...	...	...	...	...	...	...	...
295	F108	Y	F	M	IKOT INUEN	MAIZE	R	2023-10-26	6	100	...	1	2	2100	2	1	1	211	4	Average-rate Buyers	1
296	F111	Y	F	M	IKOT INUEN	BEANS	W	2023-10-26	9	30000	...	1	2	273000	2	1	4	214	7	High-rate Buyers	1
297	F111	Y	F	M	IKOT INUEN	PLANTAIN	R	2023-10-26	6	500	...	1	2	273000	2	1	4	214	7	High-rate Buyers	1
298	F109	Y	F	M	IKOT INUEN	PLANTAIN	W	2023-10-26	8	3000	...	1	1	24000	2	1	2	212	5	Average-rate Buyers	1
299	F110	Y	F	M	IKOT INUEN	COCO YAM	R	2023-10-26	9	500	...	1	1	4500	2	1	1	211	4	Average-rate Buyers	1

98 rows × 22 columns

Table 7. RFM model with Cluster 2 Result

	RESPID	AGE	GENDER	MSTATUS	VILLAGE	PRODUCE	TYPE	LASTPURCHASE	QUANTITY	PRICE	...	RECENCY	FREQUENCY	MONETARY	R	F	M	RFM_Segment	RFM_Score	General_Segment	Cluster
0	F002	O	M	M	IKOT INUEN	PALM OIL	W	2023-10-25	10	20000	...	2	3	285000	1	2	4	124	7	High-rate Buyers	2
1	F002	O	M	M	IKOT INUEN	PALM OIL	R	2023-10-25	10	500	...	2	3	285000	1	2	4	124	7	High-rate Buyers	2
2	F002	O	M	M	IKOT INUEN	PALM KERNEL	W	2023-10-25	8	10000	...	2	3	285000	1	2	4	124	7	High-rate Buyers	2
6	F005	O	M	M	IKOT INUEN	PALM OIL	W	2023-10-25	8	20000	...	2	4	260250	1	4	4	144	9	High-rate Buyers	2
7	F005	O	M	M	IKOT INUEN	BEANS	R	2023-10-25	5	250	...	2	4	260250	1	4	4	144	9	High-rate Buyers	2
...	...	...	...	...	...	...	...	...	...	...	...	...	...	...	...	...	...	...	...	...	...
259	F092	Y	F	M	OBIO AKPA	COWPEA	R	2023-10-26	7	250	...	0	4	347250	2	4	4	244	10	High-rate Buyers	2
284	F105	Y	F	M	IKOT OKORO	MAIZE	W	2023-10-26	5	5000	...	1	4	203000	2	4	3	243	9	High-rate Buyers	2
285	F105	Y	F	M	IKOT OKORO	PLANTAIN	R	2023-10-26	6	500	...	1	4	203000	2	4	3	243	9	High-rate Buyers	2
286	F105	Y	F	M	IKOT OKORO	BANANA	W	2023-10-26	10	2500	...	1	4	203000	2	4	3	243	9	High-rate Buyers	2
287	F105	Y	F	M	IKOT OKORO	COCO YAM	W	2023-10-26	10	15000	...	1	4	203000	2	4	3	243	9	High-rate Buyers	2

92 rows × 22 columns

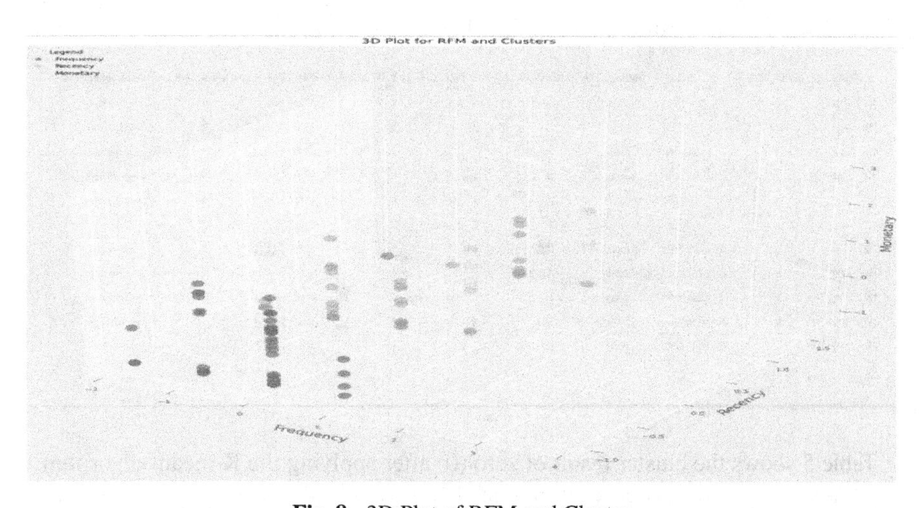

Fig. 8. 3D Plot of RFM and Cluster

4 Multiple Criteria-Based Customer Segmentation and Recommender Framework for Rural Farmers

Data gathered from farmers residing in different villages (Offline Data) undergoes a machine learning algorithm (K-mean) to develop a customer segmentation algorithm model. Subsequently, the Association rule-based apriori algorithm is employed on the customer segmentation framework shown in Fig. 9 to provide recommendations for agricultural produce.

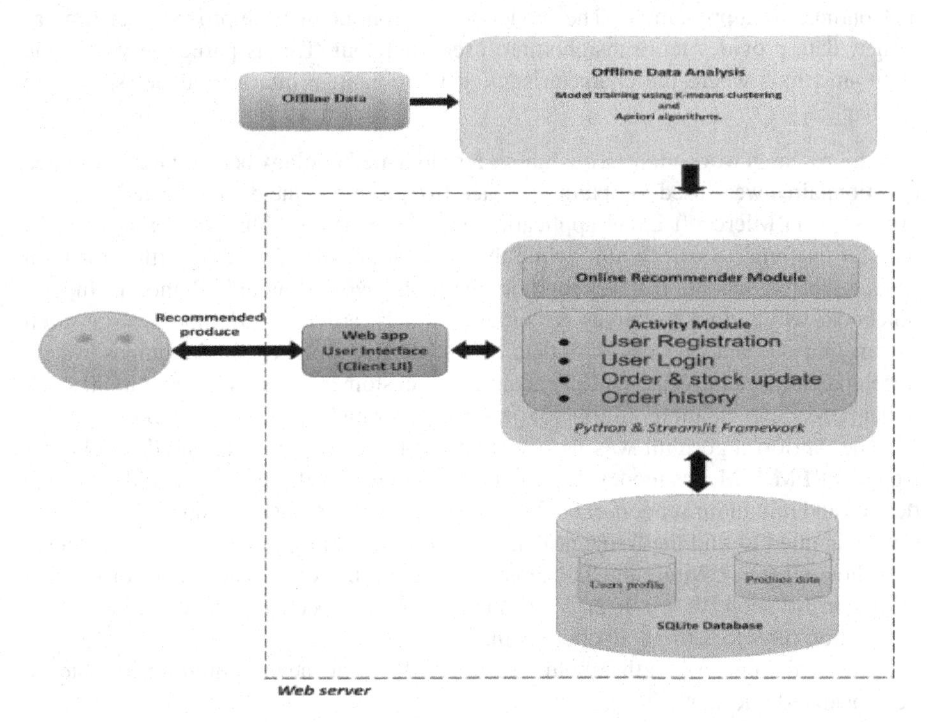

Fig. 9. Proposed System Architecture

The analysis of offline data shown in Fig. 9 is then transmitted to an online recommender model hosted on the web server for real-time recommendations. Online recommendation, also known as online learning or real-time recommendation, is an approach in recommender systems where the model is continually updated with new data. Unlike offline model training, which relies on a fixed dataset and trains the model in batches, online recommendation systems dynamically adapt to evolving user preferences by updating the model in real-time or near real-time. The activity model module manages user registration, user login, order and stock updates, as well as order history, which are stored and retrieved from the Sqlite database. The client UI serves as the user interface, and the machine learning recommendations can be directly provided from a list generated by the online recommender algorithm.

Key features of online recommendation systems include:

1. Incremental Learning: Online recommendation systems update the model parameters incrementally as new data arrives. This allows the model to adapt quickly to changing user behaviour.
2. Real-time Responsiveness: The system can provide recommendations in real-time or with low latency, making it suitable for dynamic environments where user preferences evolve rapidly.
3. Adaptability: Online recommendation systems are well-suited for scenarios where user interactions and preferences change frequently. The model can adapt to these changes without requiring a complete retraining process.
4. Continuous Improvement: The model can be continuously improved over time as new data provides more insights into user behaviour. This is particularly valuable in situations where the characteristics of the user-item interactions are subject to change.

This research work adopts a qualitative research methodology because well-designed questionnaires were used to obtain feedback from farmers, the data collected was pre-processed on Microsoft Excel application and K-means algorithm was applied to the data for customer segmentation, while the association rule apriori algorithm for farm produce recommendation. The programming tools involved include Python in Jupyter notebook, while other resources are online journals, articles and other books as well as carrying out analyses on those data collected. The Timebox model guides system development. The implementation encompasses customer and seller registration, stock updates, recommendation/ordering of produce, order history, and order reports.

The Apriori algorithm was used to find the frequent item sets based on clusters from the RFM-K Means model. The frequent itemsets for different clusters (0,1,2) were derived and minimum support of 0.03 for cluster 2, 0.02 for cluster 1 and 0.01 for cluster 0 were applied to find the frequent itemsets(The algorithm grouped buyers and sellers into three clusters). Minimum threshold of 1, minimum confidence score of 0.7 and minimum lift score of =>4, 2, 2 for cluster 2, 1, 0 respectively were applied to the association rules to achieve a better result.

The minimum support threshold determines the frequency required for a cluster to be considered relevant.

Confidence in this represents the reliability of the association rules. A higher confidence threshold means more reliable recommendations.

It is recognized how much influence and impact the agricultural sector has on the present economy; whereas agricultural marketing brings producers and consumers together through a series of activities and thus becomes an essential element of the economy. The effect of access and poor sales competition in the rural areas of farm production often led to difficulty in sales of these farm produce. The rural farmers are often burdened with the challenges of prompt sales of their harvest, especially perishable products that barely get the attention of targeted consumers. The ripple effect of this cause leads to a short availability and distribution of certain farm crops across the nation, as the issue of scarcity may also arise from the concern of high cost and inflation. Due to this concern, an emergent need to tackle these challenges is required, not only to make farmers accessible to this storage infrastructure but to also and most importantly provide a platform that handles promising connection for sales of farm produce from the rural areas, modernly connected to buyers, and ease of access and marketing awareness. At the end of this research work, the developed system has contributed to knowledge by allowing the customers (buyers) and farmers (sellers) to carry out transactions, which in turn has increased the sales of farm produce and ensured ease of access and marketing awareness. The RFM model with K-means clustering and Apriori algorithms developed has helped in effective and adequate segmentation of customers and recommendation of farm produce on web platforms. The web-based application has helped the consumers and farmers to register, login, carry out operations like getting recommended produce, ordering of produce, viewing of ordering report, updating of stock and viewing of ordering history. The application and usage of this system can be seen in the agricultural sector and the ministry of agriculture to educate and spread more awareness to subsistence and commercial farmers.

5 Results and Discussion

This focuses on system requirements, justification of programming languages, system implementation, system testing, method of system conversion, evaluation of results and discussion of results.

Implementation Procedure
This section describes the realization of the proposed system. The expected outcome from the implementation of the various design objectives is explained below:

a) ***Register Customer:*** The snapshot in Fig. 4.3 shows the system as it allows customers to register, and the information saved to the database.
b) ***Register Seller:*** The snapshot in Fig. 4.4 shows the system as it allows sellers to register, and the information saved to the database.
c) ***Login Customer:*** The snapshot in Fig. 4.5 shows the system as it allows the customers to have access to the application by entering correct login credentials.
d) ***Login Seller:*** The snapshot in Fig. 4.6 shows the system as it allows the sellers to have access to the application by entering correct login credentials.
e) ***Stock:*** This module allows the sellers to input information of different produce and update the stock of produce. The snapshot in Fig. 4.7 shows the design.
f) ***Recommend/Order produce*****:** The snapshot in Fig. 48 shows the system as it enables the customers to get recommendations on produce and to order different products.
g) ***Order history*****:** Shows the system as it allows the customer to view order reports.
h) ***Order report:*** This is the module that displays information about the order history of the customer.

6 Apriori Algorithm based on K-means Clustering Result

The Apriori algorithm was used to find the frequent item sets based on clusters from the RFM-K Means model. The frequent itemsets for different clusters (0,1,2) were derived and minimum support of 0.03 for cluster 2, 0.02 for cluster 1 and 0.01 for cluster 0 were applied to find the frequent itemsets. Minimum threshold of 1, minimum confidence score of 0.7 and minimum lift score of =>4, 2, 2 for cluster 2, 1, 0 respectively were applied to the association rules to achieve a better result.

Table 8, 9 and 10 show the result of the model for the three clusters.

Table 8. Result of the model for clusters 0

	antecedents	consequents	antecedent support	consequent support	support	confidence	lift	leverage	conviction	zhangs_metric
17	(YAM)	(COWPEA)	0.04878	0.243902	0.04878	1.0	4.100000	0.036883	inf	0.794872
30	(BANANA, CASSAVA)	(BEANS)	0.02439	0.487805	0.02439	1.0	2.050000	0.012493	inf	0.525000
36	(BANANA, MAIZE)	(BEANS)	0.02439	0.487805	0.02439	1.0	2.050000	0.012493	inf	0.525000
42	(BANANA, WHEAT)	(BEANS)	0.02439	0.487805	0.02439	1.0	2.050000	0.012493	inf	0.525000
60	(PALM OIL, CASSAVA)	(BEANS)	0.02439	0.487805	0.02439	1.0	2.050000	0.012493	inf	0.525000
62	(COCO YAM, MAIZE)	(BEANS)	0.02439	0.487805	0.02439	1.0	2.050000	0.012493	inf	0.525000
74	(COCO YAM, WHEAT)	(BEANS)	0.02439	0.487805	0.02439	1.0	2.050000	0.012493	inf	0.525000
90	(COCO YAM, CASSAVA)	(PLANTAIN)	0.02439	0.292683	0.02439	1.0	3.416667	0.017252	inf	0.725000
91	(COCO YAM, PLANTAIN)	(CASSAVA)	0.02439	0.170732	0.02439	1.0	5.857143	0.020226	inf	0.850000
92	(CASSAVA, PLANTAIN)	(COCO YAM)	0.02439	0.097561	0.02439	1.0	10.250000	0.022011	inf	0.925000
96	(COCO YAM, MAIZE)	(WHEAT)	0.02439	0.097561	0.02439	1.0	10.250000	0.022011	inf	0.925000
97	(COCO YAM, WHEAT)	(MAIZE)	0.02439	0.146341	0.02439	1.0	6.833333	0.020821	inf	0.875000
102	(MAIZE, COWPEA)	(YAM)	0.02439	0.048780	0.02439	1.0	20.500000	0.023200	inf	0.975000
103	(MAIZE, YAM)	(COWPEA)	0.02439	0.243902	0.02439	1.0	4.100000	0.018441	inf	0.775000
108	(PLANTAIN, YAM)	(COWPEA)	0.02439	0.243902	0.02439	1.0	4.100000	0.018441	inf	0.775000
114	(PALM KERNEL, MAIZE)	(PALM OIL)	0.02439	0.390244	0.02439	1.0	2.562500	0.014872	inf	0.625000
116	(MAIZE, PALM OIL)	(PALM KERNEL)	0.02439	0.097561	0.02439	1.0	10.250000	0.022011	inf	0.925000
121	(MAIZE, PLANTAIN)	(WHEAT)	0.02439	0.097561	0.02439	1.0	10.250000	0.022011	inf	0.925000
122	(WHEAT, PLANTAIN)	(MAIZE)	0.02439	0.146341	0.02439	1.0	6.833333	0.020821	inf	0.875000
127	(PALM KERNEL, PLANTAIN)	(PALM OIL)	0.02439	0.390244	0.02439	1.0	2.562500	0.014872	inf	0.625000
132	(COCO YAM, MAIZE, WHEAT)	(BEANS)	0.02439	0.487805	0.02439	1.0	2.050000	0.012493	inf	0.525000
133	(COCO YAM, MAIZE, BEANS)	(WHEAT)	0.02439	0.097561	0.02439	1.0	10.250000	0.022011	inf	0.925000
134	(COCO YAM, WHEAT, BEANS)	(MAIZE)	0.02439	0.146341	0.02439	1.0	6.833333	0.020821	inf	0.875000
135	(MAIZE, WHEAT, BEANS)	(COCO YAM)	0.02439	0.097561	0.02439	1.0	10.250000	0.022011	inf	0.925000
136	(COCO YAM, MAIZE)	(WHEAT, BEANS)	0.02439	0.048780	0.02439	1.0	20.500000	0.023200	inf	0.975000
137	(COCO YAM, WHEAT)	(MAIZE, BEANS)	0.02439	0.048780	0.02439	1.0	20.500000	0.023200	inf	0.975000

Table 9. Result of the model for clusters 1

	antecedents	consequents	antecedent support	consequent support	support	confidence	lift	leverage	conviction	zhangs_metric
42	(BANANA, CASSAVA)	(BEANS)	0.02439	0.292683	0.02439	1.0	3.416667	0.017252	inf	0.725000
44	(CASSAVA, BEANS)	(BANANA)	0.02439	0.317073	0.02439	1.0	3.153846	0.016657	inf	0.700000
59	(BANANA, COWPEA)	(COCO YAM)	0.02439	0.121951	0.02439	1.0	8.200000	0.021416	inf	0.900000
60	(COCO YAM, COWPEA)	(BANANA)	0.02439	0.317073	0.02439	1.0	3.153846	0.016657	inf	0.700000
66	(COCO YAM, YAM)	(BANANA)	0.04878	0.317073	0.04878	1.0	3.153846	0.033314	inf	0.717949
72	(PALM KERNEL, PALM OIL)	(BANANA)	0.02439	0.317073	0.02439	1.0	3.153846	0.016657	inf	0.700000
84	(PALM KERNEL, YAM)	(BANANA)	0.02439	0.317073	0.02439	1.0	3.153846	0.016657	inf	0.700000
89	(PALM OIL, YAM)	(BANANA)	0.02439	0.317073	0.02439	1.0	3.153846	0.016657	inf	0.700000
94	(WHEAT, YAM)	(BANANA)	0.02439	0.317073	0.02439	1.0	3.153846	0.016657	inf	0.700000
98	(WHEAT, COWPEA)	(BEANS)	0.02439	0.292683	0.02439	1.0	3.416667	0.017252	inf	0.725000
99	(WHEAT, BEANS)	(COWPEA)	0.02439	0.121951	0.02439	1.0	8.200000	0.021416	inf	0.900000
104	(COWPEA, YAM)	(BEANS)	0.02439	0.292683	0.02439	1.0	3.416667	0.017252	inf	0.725000
106	(BEANS, YAM)	(COWPEA)	0.02439	0.121951	0.02439	1.0	8.200000	0.021416	inf	0.900000
113	(PALM KERNEL, COWPEA)	(PLANTAIN)	0.02439	0.317073	0.02439	1.0	3.153846	0.016657	inf	0.700000
114	(PLANTAIN, COWPEA)	(PALM KERNEL)	0.02439	0.073171	0.02439	1.0	13.666667	0.022606	inf	0.950000
118	(PALM KERNEL, PALM OIL)	(YAM)	0.02439	0.195122	0.02439	1.0	5.125000	0.019631	inf	0.825000
119	(PALM KERNEL, YAM)	(PALM OIL)	0.02439	0.414634	0.02439	1.0	2.411765	0.014277	inf	0.600000
120	(PALM OIL, YAM)	(PALM KERNEL)	0.02439	0.073171	0.02439	1.0	13.666667	0.022606	inf	0.950000
130	(BANANA, PALM OIL, PALM KERNEL)	(YAM)	0.02439	0.195122	0.02439	1.0	5.125000	0.019631	inf	0.825000
131	(BANANA, PALM OIL, YAM)	(PALM KERNEL)	0.02439	0.073171	0.02439	1.0	13.666667	0.022606	inf	0.950000
132	(BANANA, PALM KERNEL, YAM)	(PALM OIL)	0.02439	0.414634	0.02439	1.0	2.411765	0.014277	inf	0.600000
133	(PALM KERNEL, PALM OIL, YAM)	(BANANA)	0.02439	0.317073	0.02439	1.0	3.153846	0.016657	inf	0.700000
137	(PALM KERNEL, PALM OIL)	(BANANA, YAM)	0.02439	0.097561	0.02439	1.0	10.250000	0.022011	inf	0.925000
138	(PALM OIL, YAM)	(BANANA, PALM KERNEL)	0.02439	0.048780	0.02439	1.0	20.500000	0.023200	inf	0.975000
139	(PALM KERNEL, YAM)	(BANANA, PALM OIL)	0.02439	0.073171	0.02439	1.0	13.666667	0.022606	inf	0.950000

Table 10. Result of the model for clusters 2

	antecedents	consequents	antecedent support	consequent support	support	confidence	lift	leverage	conviction	zhangs_metric
275	(BANANA, PALM OIL, COWPEA)	(COCO YAM)	0.045455	0.181818	0.045455	1.0	5.500000	0.037190	inf	0.857143
281	(COCO YAM, PALM OIL)	(BANANA, COWPEA)	0.045455	0.090909	0.045455	1.0	11.000000	0.041322	inf	0.952381
289	(BANANA, MAIZE, PLANTAIN)	(COCO YAM)	0.045455	0.181818	0.045455	1.0	5.500000	0.037190	inf	0.857143
295	(COCO YAM, MAIZE)	(BANANA, PLANTAIN)	0.045455	0.181818	0.045455	1.0	5.500000	0.037190	inf	0.857143
297	(COCO YAM, PLANTAIN)	(BANANA, MAIZE)	0.045455	0.090909	0.045455	1.0	11.000000	0.041322	inf	0.952381
304	(COCO YAM, COWPEA, BEANS)	(CASSAVA)	0.045455	0.181818	0.045455	1.0	5.500000	0.037190	inf	0.857143
350	(PALM OIL, CASSAVA)	(MAIZE, BEANS)	0.045455	0.227273	0.045455	1.0	4.400000	0.035124	inf	0.809524
480	(MAIZE, PLANTAIN, COWPEA)	(CASSAVA)	0.045455	0.181818	0.045455	1.0	5.500000	0.037190	inf	0.857143
494	(MAIZE, COWPEA, PALM OIL)	(WHEAT)	0.045455	0.181818	0.045455	1.0	5.500000	0.037190	inf	0.857143
496	(MAIZE, WHEAT)	(PALM OIL, COWPEA)	0.045455	0.181818	0.045455	1.0	5.500000	0.037190	inf	0.857143
509	(PLANTAIN, MAIZE, COWPEA, BEANS)	(CASSAVA)	0.045455	0.181818	0.045455	1.0	5.500000	0.037190	inf	0.857143
515	(PLANTAIN, MAIZE, COWPEA)	(CASSAVA, BEANS)	0.045455	0.181818	0.045455	1.0	5.500000	0.037190	inf	0.857143
518	(PLANTAIN, CASSAVA, COWPEA)	(MAIZE, BEANS)	0.045455	0.227273	0.045455	1.0	4.400000	0.035124	inf	0.809524
540	(MAIZE, COWPEA, BEANS, PALM OIL)	(WHEAT)	0.045455	0.181818	0.045455	1.0	5.500000	0.037190	inf	0.857143
542	(MAIZE, WHEAT, BEANS)	(PALM OIL, COWPEA)	0.045455	0.181818	0.045455	1.0	5.500000	0.037190	inf	0.857143
545	(PALM OIL, WHEAT, COWPEA)	(MAIZE, BEANS)	0.045455	0.227273	0.045455	1.0	4.400000	0.035124	inf	0.809524
548	(MAIZE, COWPEA, PALM OIL)	(WHEAT, BEANS)	0.045455	0.136364	0.045455	1.0	7.333333	0.039256	inf	0.904762
551	(MAIZE, WHEAT)	(PALM OIL, COWPEA, BEANS)	0.045455	0.136364	0.045455	1.0	7.333333	0.039256	inf	0.904762

7 Conclusion

Agriculture plays an important role in our economy; whereas agricultural marketing brings producers and consumers together through a series of activities and thus becomes an essential element of the economy. The effect of access and poor sales competition in the rural areas of farm production often led to difficulty in sales of these farm produce. The rural farmers are often burdened with the challenges of prompt sales of their harvest, especially perishable products that barely get the attention of targeted consumers. The ripple effect of this cause leads to a short availability and distribution of certain farm crops across the nation, as the issue of scarcity may also arise from the concern of high cost and inflation. Due to this concern, an emergent need to tackle these challenges is required, not only to make farmers accessible to this storage infrastructure but to also and most importantly provide a platform that handles promising connection for sales of farm produce from the rural areas, modernly connected to buyers, and ease of access and marketing awareness.

Conclusively, the developed system has contributed to knowledge by allowing the customers (buyers) and farmers (sellers) to carry out transactions, which in turn has increased the sales of farm produce and ensured ease of access and marketing awareness. The RFM model with K-means clustering and Apriori algorithms developed has helped in effective and adequate segmentation of customers and recommendation of farm produce on web platforms. The web-based application has helped the consumers and farmers to register, login, carry out operations like getting recommended produce, ordering of produce, viewing of ordering report, updating of stock and viewing of ordering history. The application and usage of this system can be seen in the agricultural sector and the ministry of agriculture to educate and spread more awareness to subsistence and commercial farmers.

Acknowledgement. The authors are grateful to TETFund for supporting this research through the TETFund Centre of Excellence in Computational Intelligence Research and the University of Uyo Management for creating a conducive environment for conducting the research.

References

1. Majumdar, J., Naraseeyappa, S., Ankalaki, S.: Analysis of agriculture data using data mining techniques: application of big data. J. Big Data **4**(1), 1–15 (2017). https://doi.org/10.1186/s40537-017-0077-4
2. Srivastava, S.K.: *Agricultural Marketing: Concept and Definitions*. Notes prepared for Course AgEcon530 (Agricultural Marketing), pp. 1–29. NCAP, New Delhi (2015)
3. Usip, P.U., Udo, E.N., Asuquo, D.E., James, O.R.: A machine learning-based mobile chatbot for crop farmers. In: Ortiz-Rodríguez, F., Tiwari, S., Sicilia, M.-A., Nikiforova, A. (eds.) Electronic Governance with Emerging Technologies: First International Conference, EGETC 2022, Tampico, Mexico, September 12–14, 2022, Revised Selected Papers, pp. 192–211. Springer Nature Switzerland, Cham (2022). https://doi.org/10.1007/978-3-031-22950-3_15
4. Usip, P.U., James, O.R., Asuquo, D.E., Udo, E.N., Osang, F.B.: Usability evaluation of an intelligent mobile chatbot for crop farmers. In: Ortiz-Rodríguez, F., Tiwari, S., Usoro Usip, P., Palma, R. (eds.) Electronic Governance with Emerging Technologies: Second International Conference, EGETC 2023, Poznan, Poland, 11–12 Sep 2023, Revised Selected Papers, pp. 101–111. Springer Nature Switzerland, Cham (2023). https://doi.org/10.1007/978-3-031-43940-7_9
5. Bezhovski, Z.: The future of mobile payment as an electronic payment system. Eur. J. Bus. Manag. **8**(8), 127–128 (2016)
6. Udokwu, C., Zimmermann, R., Darbanian, F., Obinwanne, T., Brandtner, P.: Design and Implementation of a Product Recommendation System with Association and Clustering Algorithms. CENTERIS – International Conference on ENTERprise Information Systems/ProjMAN – International Conference on Project MANagement/HCist – International Conference on Health and Social Care Information Systems and Technologies 2022. Procedia Comput. Sci. **219**, 512–520 (2023)
7. Gatkal, N.R., Nalawade, S.M., Khurdal, J.K., Pawase, P.P.: Machine learning in agriculture. Just Agric. Multi. e-Newsletter **3**(7), 168 (2023)
8. Shi, Z., Wen, Z., Xia, J.: An Intelligent Recommendation system based on customer segmentation. Int. J. Res. Bus. Stud. Manag. **2**(11), 78–90 (2015)
9. Kamble, N., Shaha, R.A.: Product Recommendation System Using Machine Learning. Department of Computer Science and Engineering, Shreeyash College of Engineering and Technology, Aurangabad, India (2022). https://ssrn.com/abstract=4245401
10. Gao, W., Zhou, Z.-H.: On the doubt about margin explanation of boosting. Artific. Intell. **203**, 1–18 (2013). https://doi.org/10.1016/j.artint.2013.07.002

Enhancing Turkish Music Emotion Prediction: A Comparative Analysis of Machine Learning Techniques

Siva Sai Susmitha Katta[1], Siva Kumar Katta[2], Junali Jasmine Jena[1](✉), Mahendra Kumar Gourisaria[1], and Suresh Chandra Satapathy[1]

[1] School of Computer Engineering, KIIT University, Bhubaneswar, India
junali.jenafcs@kiit.ac.in

[2] Department Computer Science, Arizona State University, Tempe, AZ, USA

Abstract. This research paper explores the classification of turkish music based on emotional content using machine learning techniques. Utilizing a dataset comprising 400 music samples evenly distributed across four emotional categories—happy, sad, angry, and relax—each sample is represented by 50 acoustic features derived from diverse genres of Turkish music. Initial experiments with individual machine learning models demonstrated promising results with accuracy of 92.5% by Gradient Boost Classifier. To enhance classification performance, this study utilizes multi-modal classifiers through stacking and voting methods, leveraging the strengths of various models. Stacking and Voting classifiers achieved an accuracy of 93% on the dataset which indicate that multi-model classifiers can achieve superior accuracy and robustness compared to individual models, offering significant implications for applications in personalized music recommendation systems, therapeutic tools, and interactive entertainment systems. This paper comprehensively evaluates individual and multi-model classifiers, detailing their methodologies, experimental results, and potential for future advancements in music emotion recognition.

Keywords: Music Emotion Recognition(MER) · Hyper-parameter Tuning · ML models · Stacking Classifier · Voting Classifier

1 Introduction

Music's ability to evoke emotions has long been recognized. Music emotion recognition (MER) has become an important research area within affective computing and machine learning in recent years. This burgeoning field explores how to quantify and classify emotional responses elicited by music [1]. Researchers have developed applications with profound implications by understanding how music influences emotions. These applications could include personalized music recommendation systems, therapeutic tools for emotional well-being, and interactive entertainment experiences that adapt to the user's emotional state [1].

S. Tiwari et al. (Eds.): AI4S 2024, CCIS 2243, pp. 43–52, 2025.
https://doi.org/10.1007/978-3-031-81369-6_4

Traditional methods for music emotion prediction, such as manual annotation, rule-based systems, and signal processing techniques, have drawbacks, including subjectivity, limited flexibility, and scalability issues. Early Music Emotion Recognition (MER) work focused on extracting low-level acoustic features from audio signals. For example, Tzanetakis and Cook (2002) [2] introduced a system for genre classification using features like timbre, rhythm, and pitch, demonstrating that emotions could be distinguished using these properties. Machine learning models address these issues by capturing complex patterns and relationships between musical features and emotions, offering scalability, consistency, adaptability, and better generalization across genres. This study proposes using a multi-model classifier to enhance the classification accuracy. The diversity in musical genres is intended to provide a robust foundation for developing and evaluating classification models.

Following sections could be found in rest of the paper: Sect. 2 presents an overview of related research in Music Emotion Recognition (MER). Section 3 details the dataset and the preprocessing steps involved. Section 4 outlines the methodologies used for both individual and multi-model classifiers. Experimention results and analysis is described in Sect. 5. Conclusion and Future work is provided in Sect. 6. By assessing individual and multi-model classifiers using this dataset, the study seeks to advance the field of MER, offering valuable insights and tools for applications in music recommendation, therapy, and interactive entertainment.

2 Literature Review

Several works have been done in the field of music genre classification. Some of the recent works have been reviewed in this section. Panda et al. [3] outlined a method for classifying emotions in music using a dataset of 903 clips from the All music database. By extracting features with Marsyas, MIR Toolbox, and Psysound frameworks and employing SVM classifiers, the work showed competitive results with F1 score of 47.2% through 20 repetitions of 10-fold cross-validation, emphasizing the importance of feature selection. Er et al. [4] described a method for predicting emotions in Turkish music using preprocessing, feature extraction, and classification via SVM, K-NN, and ANN. After noise removal, normalizing sampling frequency, and feature vector extraction, 79.3% accuracy was achieved by the model on a four-class dataset, demonstrating its effectiveness. Durahim et al. [5] developed a model for emotion recognition using song lyrics and machine learning. By annotating 300 songs, text preprocessing, and extracting Unigram, Bigram, and Trigram features, the study achieves a recall of 43.7 and precision of 46.9 by Multinomial Naïve Bayes. Hizlisoy et al. [6] introduced a novel approach using a CLDNN architecture on a Turkish emotional music database. Incorporating log-mel filterbank energies and MFCCs with standard acoustic features, the model achieves 99.19% accuracy through 10-fold cross-validation, outperforming k-NN, SVM, and Random Forest classifiers. Er et al. [7] presented a new approach combining pre-trained deep learning models, such as AlexNet and VGG-16, with chroma spectrograms from music recordings. By extracting deep visual characteristics from various layers of these models, the VGG-16 model achieves 89.2% accuracy in the Fc7 layer, highlighting the effectiveness of deep learning in capturing emotional elements in music. J. Song [8] compared the performance of these models on the Turkish Music Emotion Dataset. The traditional random forest model outperforms XGBoost with 80.8% accuracy, an 80.8% recall rate, and an 80.5% F1 score. Ezzaidi et al. [9] trained statistical pattern recognition classifiers using real-world audio collections for ten genres and achieved 61% accuracy. The paper offers a framework for creating characteristics for content-based analysis of musical signals and highlights the possibilities of automatic genre classification. Kolacoglu et al. [10] outlined a method for recognizing emotions from speech in multiple languages. By preprocessing data, extracting features with OpenSMILE, and using various classification algorithms, including logistic regression, the method achieves recognition rates of 92.73% for the Turkish dataset and 96.3% for the Emo-db dataset. Cimtay et al. [11] proposed a multi-modal method using facial expressions, GSR, and EEG for emotion recognition. It attains a mean accuracy of 74.2% for three emotion classes on the LUMED-2 dataset and 91.5% and 53.8%, respectively, for different emotion classes on the DEAP dataset. Er et al. [12] introduced an EEG-based emotion recognition model using deep learning for music listening. The method achieves notable performance by recording EEG signals, extracting spectrograms, and using pre-trained deep network models like AlexNet and VGG16, which achieves notable performance, with VGG16 achieving the best result. Costa et al. [13] presented a multi-modal approach combining audio and textural features for emotion recognition on a 1500 song dataset and the study achieves its best performance with 44.2% using audio features, 46.3% with textural features under supervised learning, and 51.3% with semi-supervised learning. Yang et al. [14] used CNN for music emotion prediction by deriving a spectogram via constant Q-transform and then applying CNN on it. The model achieved an accuracy of 82.24% and 81.80% in the valence and arousal dimension respectively.

3 Materials and Methods

3.1 Dataset Description

The turkish music emotion dataset [15] is a discrete model with four classes: happy, sad, angry, and relax. It consists of 400 samples, each 30 s long, selected from different genres of Turkish music encompassing both verbal and non-verbal forms. The number of samples in each class equals 100 samples for each class. The target variable is categorical, with class labels being "relax," "happy," "sad," and "angry" and the data distribution is given in Fig. 1.

To capture the music's essential elements often associated with human emotional responses, 50 acoustic features were extracted from each sample. These features include:

- RMS Energy: Measures the audio signal's root mean square, representing the music's power or loudness.
- Spectral Centroid: The Spectral Centroid Represents the center of mass of the spectrum, typically linked to the perceived brightness of a sound.
- Mel-Frequency Cepstral Coefficients (MFCCs): Captures the timbral aspects of the audio, which are widely used in audio processing and music classification.

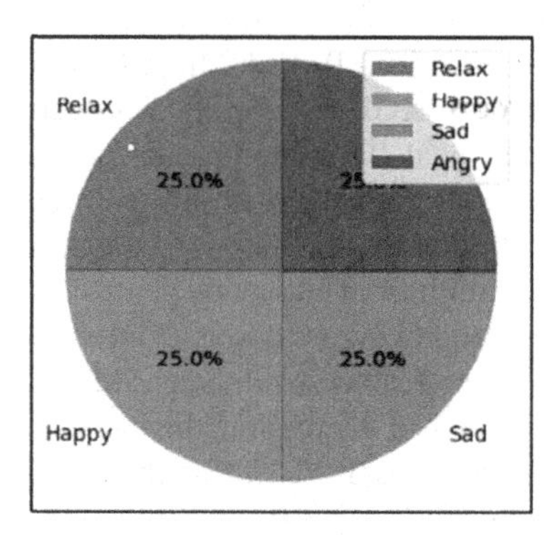

Fig. 1. Data distribution in Turkish Music Emotion Prediction Dataset

3.2 Methodology

The dataset offers unique challenges, allowing for a multifaceted analysis leveraging machine learning (ML) to enhance the accuracy of emotion prediction from music. This section outlines the uniform steps and procedures undertaken for the research work. The dataset's standardization boosted accuracy while normalization decreased it, leading us to opt for standardization in our model.

Models Used

- Individual Classifiers

Logistic Regression [16]- logistic regression tweaks a linear equation separating data points into classes.

Random Forest [17]- Random Forest generates multiple decision trees from your data. Support Vector Machine (SVM) [18]- SVMs find a dividing line (or hyperplane in higher dimensions) that maximizes the margin between the classes.

Naive Bayes [19]- Naive Bayes computes the likelihood of a data point belonging to a particular class by considering each feature independently.

K Nearest Neighbors (KNN) [20]- KNN memorizes your training data. When a new data point arrives, it finds the nearest K data points. here number of neighbors is considered as 5.

Decision Tree [21]- Decision Trees split data based on feature values, creating branches for possible outcomes until reaching a decision.

Gradient Boosting [22]- Gradient Boosting builds sequential decision trees, each correcting errors of the previous. Here number of estimators is considered as 5000.

AdaBoost [23]- AdaBoost combines weak classifiers, adjusting misclassified instance weights for better focus. Here Base estimator is DecisionTreeClassifier with max_depth = 1 and number of estimators is 5000.

CatBoost [24]- CatBoost handles categorical features directly and uses ordered boosting to reduce overfitting. Here, Iterations is taken as 100, Learning rate as 0.1, and Depth as 6.

Multi-layer Perceptron Classifier (MLP) [25]- MLP is an artificial neural network with multiple layers, learning by adjusting connection weights via backpropagation. Here no. of Hidden layers is taken as 1 with 100 neurons and maximum iterations as 500.

Hyperparameter tuning was conducted using GridSearchCV [26] to optimize the performance of the best models. The models tuned were Random Forest, Logistic Regression, and Gradient Boosting Classifier.

Random Forest Hyperparameter Tuning-

n_estimators: [100, 200, 500], max_depth: [None, 10, 20, 30], min_samples_split: [1, 4, 9], min_samples_leaf: [1, 3].

Logistic Regression Hyperparameter Tuning-

Solver: ['newton-cg,' 'lbfgs,' 'liblinear'].

Gradient Boosting Classifier Hyperparameter Tuning-

n_estimators: [100, 200, 500], learning_rate: [0.01, 0.1, 0.2], max_depth: [2, 4, 6].

- **Multi-model Classifiers**

Multi-model classifiers were employed to enhance classification accuracy further using techniques such as stacking and voting. These approaches leverage the strengths of multiple models to achieve better performance.

Stacking classifier: It entails training several base models and employing a meta-model to aggregate their predictions [27].

-Base Models: Logistic Regression, Random Forest, SVM.

-Meta-Model: Logistic Regression.

Voting Classifier- Voting involves training multiple models and combining their predictions through a majority vote (for classification) or averaging (for regression) [27]. Soft voting, which averages the predicted probabilities, was used in this study.

-Base Models: Logistic Regression, Random Forest, SVM.

Principal Component Analysis (PCA) has been used with variance of 95% for reducing feature set [27].

Model Evaluation

Evaluation metrics include Accuracy Score, Confusion Matrix (TP, TN, FP, FN), ROC Curve, AUC, and Classification Report (Precision, Recall, F1-Score, Support) to assess model performance.

4 Results and Discussion

Experimentation details along with its appropriate analysis is provided in this section following the methodologies described in Sect. 4. All the simulations have been carried out using Windows 11 Operating system having AMD Ryzen 7 5700U processor with

Radeon Graphics and 16GB RAM and analyzed using Python 3 with a Jupyter notebook. Various individual and multi-model classifiers were evaluated to determine their effectiveness in classifying music into the four emotional categories: happy, sad, angry, and relax. The performance of each model is discussed, and relevant plots and pictures are provided to illustrate the results. Table 1 provides the training accuracy and testing accuracy of the models used. Table 2 provides values of Precision and Recall for the ML models categorized into four classes.

Table 1. Accuracy of machine learning models.

Model	Train Accuracy	Test Accuracy
LR	0.86	0.982
RF	0.975	0.938
SVM	0.91	0.973
NB	0.79	0.921
KNN	0.78	0.956
DT	1.0	0.71
GBM	1.0	0.925
ADA	0.88	0.84
CAT	0.99	0.875
MLP	0.97	0.8

Table 2. Precision and Recall values of ML models.

Model	Precision				Recall			
	Relax	Happy	Sad	Angry	Relax	Happy	Sad	Angry
LR	1.00	0.85	0.83	0.95	0.85	0.96	0.83	0.95
RF	0.76	0.88	0.82	0.80	0.95	0.91	0.50	0.84
SVM	0.85	0.79	0.79	1.00	0.85	0.96	0.61	0.95
NB	0.86	0.81	0.83	0.86	0.90	0.91	0.56	0.95
KNN	0.70	0.66	0.69	0.93	0.70	0.91	0.50	0.74
DT	0.62	0.84	0.52	0.94	0.75	0.70	0.61	0.79
GBM	0.90	0.92	0.88	1.00	0.90	0.96	0.83	1.00
ADA	0.80	0.95	0.68	0.90	0.60	0.91	0.83	1.00
CAT	0.94	0.92	0.70	0.95	0.75	0.96	0.78	1.00
MLP	0.84	0.88	0.76	0.95	0.80	0.96	0.72	0.95

From the result analysis, the effectiveness of various machine learning models in classifying music into four emotional categories: happy, sad, angry, and relax was determined. The Gradient Boosting Machine (GBM) emerged as the top-performing individual model with an accuracy of 92.5%, closely followed by Logistic Regression and Random Forest, which achieved accuracies of 90% and 81.25%, respectively.

Hyperparameter Tuning Results

To optimize their performance, hyperparameter tuning was conducted for the best models, Random Forest, Logistic Regression, and Gradient Boosting Classifier, which is provided in previous section. The process of hyperparameter tuning significantly improved the performance of the models.

Multi-model Classifier Results

The stacking classifier, which combined the predictions of logistic regression, random forest, and SVM with a logistic regression meta-model, achieved improved performance for individual models (Training Accuracy: 0.98, Testing Accuracy: 0.93).

The voting classifier combined random forest, logistic regression, and gradient boosting using soft voting, enhancing performance compared to individual models (Training Accuracy: 1.00, Testing Accuracy: 0.93).

Table 3. Precision, Recall and F1_Score values of the stacking and voting classifier.

	Stacking Classifier			Voting Classifier		
Class	Precision	Recall	F1-score	Precision	Recall	F1-score
Relax	0.93	0.92	0.93	0.90	0.90	0.90
Happy	0.92	0.93	0.93	0.91	0.91	0.91
Sad	0.92	0.93	0.93	0.89	0.89	0.89
Angry	0.93	0.93	0.93	1.00	1.00	1.00

The stacking and voting classifiers demonstrated improved performance compared to individual models and achieved an average accuracy of 93%, indicating that combining the strengths of multiple models can lead to superior performance, which is shown in Table 3. Many researches have proved that ensemble classifiers outperforms single classifiers and it is expected too. While GBM and ensemble models showed high accuracy, simpler models like Logistic Regression also performed competitively. This suggests that while complex models can capture more intricate patterns in the data, simpler models can still provide robust performance, especially when computational efficiency is a consideration.

Table 4 provides comparison of proposed work with other similar work from the literature. From the table it could be inferred that Stacking classifier showed competent and promising results in comparison to other works.

Table 4. Comparative performance analysis of proposed methodology with other works

Sl.No	Works	Accuracy of the models(%)
1	M. B. Er and E. M. Esinl [4]	79.3
2	M. B. Er and I. B. Aydilek [7]	89.2
3	J. Song [8]	80.8
4	H.Ezzaidi & G. Essl [9]	61.0
5	Our Work	93.0

5 Conclusion

This research demonstrated the potential of machine learning models, particularly ensemble methods, in effectively classifying music based on emotional content. The insights gained from this study contribute valuable knowledge to the field of music emotion recognition and pave the way for further advancements in personalized music recommendation systems, therapeutic tools, and interactive entertainment systems. Continued innovation and interdisciplinary collaboration will be essential for addressing the remaining challenges and unlocking the full potential of music emotion recognition. This research paper explored the classification of music emotions using a dataset comprising 400 samples and various machine learning model were employed to establish a performance baseline. Among these, the Gradient Boosting Machine emerged as the top-performing individual model with an accuracy of 92.5%, closely followed by Logistic Regression and Random Forest. The study demonstrated the importance of hyperparameter tuning in enhancing model performance. Optimizing parameters such as the number of estimators, maximum depth, and learning rates significantly improved the accuracy and reliability of the models.

Despite the promising results, there are several areas for future research, such as lyrical content and metadata can be taken into account, exploring CNN and RNN techniques, integrating multi-modal data, and leveraging unsupervised and semi-supervised learning methods could further enhance the performance of music emotion recognition systems. Additionally, developing real-time emotion recognition capabilities and expanding the dataset to include diverse cultural backgrounds are potential directions for future work.

References

1. Yang, Y.-H., Lin, Y.-C., Su, Y.-F., Chen, H.H.: A regression approach to music emotion recognition. IEEE Trans. Audio Speech Lang. Process. **16**(2), 448–457 (2008). https://doi.org/10.1109/TASL.2007.911513
2. Oramas, S., Nieto, O., Barbieri, F., Serra, X.: Multi-label music genre classification from audio, text, and images using deep features. arXiv preprint arXiv:1707.04916 (2018)
3. Panda, R., Paiva, R.P.: Music emotion classification: Dataset acquisition and comparative analysis. In: 15th International Conference on Digital Audio Effects–DAFx (2012)
4. Er, M.B., Esin, E.M.: Music emotion recognition with machine learning based on audio features. Comput. Sci. **6**(3), 133–144 (2021)

5. Durahim, A.O., Setirek, A.C., Özel, B.B., Kebapçı, H.: Music emotion classification for Turkish songs using lyrics. Pamukkale J. Eng. Sci. **24**(2), 292–301 (2018). https://doi.org/10.5505/pajes.2017.15493
6. Hizlisoy, S., Yildirim, S., Tufekci, Z.: Music emotion recognition using convolutional long short term memory deep neural networks. Eng. Sci. Technol. Int. J. **24**(3), 760–767 (2021)
7. Er, M.B., Aydilek, I.B.: Music emotion recognition by using chroma spectrogram and deep visual features. Int. J. Comput. Intell. Syst. **12**(2), 1622–1634 (2020)
8. Er, M.B., Aydilek, I.B.: Music emotion recognition by using chroma spectrogram and deep visual features. Int. J. Comput. Intell. Syst. **12**(2), 1622–1634 (2023)
9. Ezzaidi, H., Rouat, J.: Automatic music genre classification using second-order statistical measures for the prescriptive approach. In: INTERSPEECH, pp. 141–144 (2005)
10. Çolakoğlu, E., Hızlısoy, S.E.R.H.A.T., Arslan, R.E.C.E.P.: Multi-lingual Speech Emotion Recognition System Using Machine Learning. Selcuk Univ. J. Eng. Sci. **23**(1), 1-11 (2024)
11. Cimtay, Y., Ekmekcioglu, E., Caglar-Ozhan, S.: Cross-subject multimodal emotion recognition based on hybrid fusion. IEEE Access **8**, 168865–168878 (2020). https://doi.org/10.1109/ACCESS.2020.3023871
12. Er, M.B., Çiğ, H., Aydilek, İB.: A new approach to recognition of human emotions using brain signals and music stimuli. Appl. Acoustics **175**, 107840 (2021)
13. Costa, Y.M.G., Oliveira, L.S., Koerich, A.L., Gouyon, F., Martins, J.G.: Music genre classification using LBP textural features. Signal Process. **92**(11), 2723–2737 (2012). https://doi.org/10.1016/j.sigpro.2012.04.023
14. Yang, P.-T., Kuang, S.-M., Wu, C.-C., Hsu, J.-L.: Predicting music emotion by using convolutional neural network. In: Fui-Hoon Nah, F., Siau, K. (eds.) HCII 2020. LNCS, vol. 12204, pp. 266–275. Springer, Cham (2020). https://doi.org/10.1007/978-3-030-50341-3_21
15. Abuchionw uegbusi, A.: Turkish Music Emotion Prediction. Kaggle Dataset. https://www.kaggle.com/datasets/abuchionwuegbusi/turkish-music-emotion-prediction (2022)
16. Schober, P., Vetter, T.R.: Logistic regression in medical research. Anesthesia Analgesia **132**(2), 365–366 (2021)
17. Genuer, R., Poggi, J.M., Genuer, R., Poggi, J.M.: Random Forests, pp. 33–55. Springer International Publishing (2020)
18. Abdullah, D.M., Abdulazeez, A.M.: Machine learning applications based on SVM classification a review. Qubahan Acad. J. **1**(2), 81–90 (2021)
19. Ampomah, E.K., Nyame, G., Qin, Z., Addo, P.C., Gyamfi, E.O., Gyan, M.: Stock market prediction with gaussian naïve bayes machine learning algorithm. Informatica **45**(2), 243–256 (2021). https://doi.org/10.31449/inf.v45i2.3407
20. Narayan, Y.: SEMG signal classification using KNN classifier with FD and TFD features. Mater. Today: Proc. **37**, 3219–3225 (2021)
21. Tangirala, S.: Evaluating the impact of GINI index and information gain on classification using decision tree classifier algorithm. Int. J. Adv. Comput. Sci. Appl. **11**(2), 612–619 (2020)
22. Khan, M.S.I., Islam, N., Uddin, J., Islam, S., Nasir, M.K.: Water quality prediction and classification based on principal component regression and gradient boosting classifier approach. J. King Saud Univ.-Comput. Inform. Sci. **34**(8), 4773–4781 (2022)
23. Zhang, L., Wang, J., An, Z.: Vehicle recognition algorithm based on Haar-like features and improved Adaboost classifier. J. Ambient Intell. Humanized Comput. **14**(2), 807–815 (2023)
24. Ogar, V.N., Hussain, S., Gamage, K.A.A.: Transmission line fault classification of multi-dataset using catboost classifier. Signals **3**(3), 468–482 (2022)
25. Pahuja, R., Kumar, A.: Sound-spectrogram based automatic bird species recognition using MLP classifier. Appl. Acoustics **180**, 108077 (2021)

26. Kartini, D., Nugrahadi, D.T., Farmadi, A.: Hyperparameter tuning using GridsearchCV on the comparison of the activation function of the ELM method to the classification of pneumonia in toddlers. In: 2021 4th International Conference of Computer and Informatics Engineering (IC2IE), pp. 390–395. IEEE (2021)
27. Zhao, R., Mu, Y., Zou, L., Wen, X.: A hybrid intrusion detection system based on feature selection and weighted stacking classifier. IEEE Access **10**, 71414–71426 (2022)
28. Uddin, M.P., Mamun, M.A., Hossain, M.A.: PCA-based feature reduction for hyperspectral remote sensing image classification. IETE Tech. Rev. **38**(4), 377–396 (2021). https://doi.org/10.1080/02564602.2020.1740615

Integrating Deep Learning and Imaging Techniques for High-Precision Brain Tumor Analysis

Dilip Kumar Gokapay(✉) and Sachi Nandan Mohanty

School of Computer Science and Engineering (SCOPE), Amaravati, Andhra Pradesh, India
{gokapay.22phd7117,sachinandan.m}@vitap.ac.in

Abstract. The brain, as the core organ of the human nervous system, governs a wide range of bodily functions. These include vital processes such as breathing and regulating heart rate, as well as higher-order functions like thinking, emotional responses, and creativity. Composed of billions of neurons and glial cells, the brain is divided into distinct regions, each specializing in specific functions. Neurons in the brain communicate through an intricate system of electrical and chemical signals, facilitating information processing, memory storage, and responses to external stimuli. Investigating brain tumors is a significant aspect of medical research that aims to enhance the accuracy of diagnoses, treatment techniques, and patient outcomes. Using cutting-edge machine learning algorithms on data collected from imaging modalities like MRI and CT scans for the purpose of tumor identification, classification, and segmentation is the main focus of this research. Combining conventional machine learning techniques with convolutional neural networks (CNNs) substantially enhances the processing and interpretation of complex medical pictures. The study also focuses on developing a comprehensive dataset, employing preprocessing techniques to enhance image clarity, and applying deep learning models to accurately detect tumor locations. We use several performance indicators to evaluate the models' efficacy, ensuring they meet clinical requirements. Our results demonstrate a significant improvement over earlier methods in brain tumor identification, with a success rate of 95.35%. This program will assist healthcare professionals, including radiologists, in making more informed decisions, which should lead to improved outcomes for patients.

Keywords: Brain Tumor · MRI Images · Convolutional Neural Networks (CNNs) · Segmentation · Deep Learning

1 Introduction

Brain tumors represent a significant medical challenge, with serious implications for diagnosis, treatment, and patient prognosis. These tumors, whether malignant or benign, arise from various cell types within the brain [1]. It is very important to quickly and accurately find brain tumors in order to improve treatment methods and patient outcomes, since they can be in tricky places and affect important brain functions. Conventional

S. Tiwari et al. (Eds.): AI4S 2024, CCIS 2243, pp. 53–67, 2025.
https://doi.org/10.1007/978-3-031-81369-6_5

approaches to the diagnosis of brain tumors mostly depend on the painstaking manual interpretation of imaging modalities like CT and MRI. It is concerning that these procedures may not be successful and that human error could lead to inaccurate diagnosis and treatment delays. It is very important to quickly and accurately find brain tumors in order to improve treatment and patient outcomes, since they can be in tricky places and affect important brain functions. Traditional approaches to detecting brain tumors using imaging modalities like CT and MRI rely significantly on human analysis. But these procedures can be laborious and error-prone, which could cause false diagnoses or treatment delays [2].

Recent advances in computing and imaging technologies have opened the door for the integration of AI and ML into medical diagnosis. In order to improve the efficiency and precision of brain tumor evaluation, this research aims to apply cutting-edge machine learning methods [3]. Using medical imaging data, we aim to build automated systems that can detect, classify, and divide brain cancers using deep learning models, particularly convolutional neural networks (CNNs). The convolutional neural network (CNN) is a deep learning system that excels in analyzing and comprehending visual data [4]. These networks are composed of several layers, including convolutional layers that autonomously learn to extract significant features from images, pooling layers that reduce the data's dimensionality while retaining critical details, and fully connected layers that handle higher-level reasoning for classification tasks.

The initial stage in using convolutional neural networks (CNNs) to analyze brain tumors is to acquire and prepare medical imaging data. It is essential to preprocess data in order to guarantee it is fit for analysis [5]. This procedure incorporates picture enhancement, noise reduction, and normalizing. The first one removes artifacts that could impact the model's performance; the second one helps maintain constant pixel intensity values; and the third one diversifies the training dataset by applying various image transformations, such as scaling, rotating, and flipping. By strengthening the model and reducing the likelihood of overfitting, these preprocessing processes guarantee that the CNN will be able to generalize effectively to novel, unknown data [6]. The architecture of the CNNs used in this study was carefully planned to strike a balance between computational complexity and efficiency. Convolutional layers can process images by applying filters that help them detect edges, textures, patterns, and more. Pooling layers subsequently downsample the feature maps, so diminishing the computational load and augmenting the invariance of the representations to minor modifications in the images [7].

In a subsequent step, fully connected layers integrate these features in order to predict the presence, categorization, and magnitude of brain tumors. Part of this job is collecting and analyzing a ton of brain images, using sophisticated preprocessing techniques, and creating strong machine learning models. We evaluate the efficacy of these models using metrics including accuracy, sensitivity, specificity, and area under the receiver operating characteristic (ROC) curve to ascertain their clinical relevance [8]. Additionally, we conduct cross-validation and external validation to assess the generalizability of our models. Hopefully, the findings of this study will improve the diagnosis and treatment of brain tumors and contribute to our current understanding of medical AI. Following this, we will describe the study's methodology, experimental design, outcomes, and comments surrounding these findings, highlighting the pros and cons of employing AI to investigate brain cancers [9]. We want to augment the functionalities of automated diagnostic tools to aid physicians in making more accurate and quicker decisions, therefore optimizing patient care and results.

Artificial intelligence poses significant ethical dilemmas and may create obstacles in the execution of medical diagnostics. Ethically, concerns arise around patient privacy, data security, and the transparency of AI decision-making processes. Disparate patient populations may experience inequitable care due to the implementation of biased AI systems arising from the utilization of unrepresentative training data. Furthermore, excessive dependency on AI may diminish the necessity for human discernment and foster reliance on technology. Real-world implementation challenges encompass the integration of AI systems with existing healthcare infrastructure, obtaining regulatory approvals, managing technology deployment costs, and addressing resistance from medical professionals apprehensive about AI's role in patient care. Finally, current machine learning methods, particularly convolutional neural networks (CNNs), show a lot of potential for improving the efficiency and accuracy of brain tumor diagnostics. Models like this can automate brain tumor identification, classification, and segmentation, improving patient outcomes and assisting clinicians in making better decisions [10].

2 Related Work

In order to detect and classify brain cancers using several imaging modalities, especially magnetic resonance imaging (MRI), ML and DL methods are essential [11–13]. With a focus on fresh methods and their outcomes, this section reviews significant and recent research in this field. Mohsen, Heba, and colleagues utilized deep learning algorithms with discrete wavelet transform (DWT) characteristics to create a hybrid system. Fuzzy c-means clustering was employed for the purpose of brain tumor segmentation [14–16]. Principal component analysis (PCA) was used to reduce the dimensionality of the variables retrieved from each diagnosed lesion before they were fed into deep neural networks (DNN). This was accomplished by means of the discrete wavelet transform (DWT). Incredible sensitivity (97.0% level) and accuracy (96.97%) were achieved by this method.

Cendy Prakarsa, Yohannes Yohannes, Widhiarso, and Wijang came up with a method for classifying brain tumors. A combination of CNNs and data extracted from the Gray Level Co-occurrence Matrix (GLCM) is used in it. Four features—energy, correlation, contrast, and homogeneity—were extracted from each image at four distinct angles: 0°, 45°, 90°, and 135°. The CNN was thereafter given the features to analyze [17–19]. They tested their approaches on four datasets: Mg-Gl, Mg-Pt, Gl-Pt, and Mg-Gl-Pt. They found that using two sets of features, contrast with homogeneity and contrast with correlation, yielded the best accuracy of 82.27% for the GPU dataset. In order to automate the process of detecting and classifying brain cancers, Seetha, J., and S. S. Raja created a method utilizing deep convolutional neural networks. After applying fuzzy C-means (FCM) to divide the brain into distinct regions, this method retrieved shape and texture information from those areas [20, 21]. After feeding the characteristics into DNN and SVM classifiers, we got a success rate of 97.5%. In order to improve the accuracy of brain tumor categorization, Cheng, Jun, and colleagues enhanced ROIs and employed precise ring-form partitioning. We enhanced the bag-of-words (BoW) approach, intensity histogram, and gray level co-occurrence matrix (GLCM) among other feature extraction techniques. Next, we trained the classifier using the feature vectors. According to the experimental data, the intensity histogram (71.39% to 78.18% improvement), GLCM (83.54% to 87.54% improvement), and BoW (91.28% improvement) all enhanced accuracy. However, the degrees of improvement varied [22–24].

Table 1. Summary of Related Work

AUTHORS	METHODOLOGY	KEY TECHNIQUES	ACCURACY
Mohsen, Heba, et al	It uses DWT features in combination with DL methods, uses fuzzy c-means for segmentation, extracts features using DWT, reduces dimensionality using PCA, and classifies using DNN	DWT, PCA, DNN	96.97%/97.0%
Widhiarso, Wijang, et al	The system uses CNN and GLCM features to derive four image-specific metrics: energy, homogeneity, correlation, and contrast	CNN, GLCM	82.27% (Gl-Pt dataset)
S. Raja and Seetha. J	Proposes deep CNN for automated detection and grading, uses FCM for segmentation, extracts texture and shape features, classifies using SVM and DNN	CNN, FCM, SVM, DNN	97.5%
Cheng, Jun, et al. [13]	Increases precision in classification by refining features like ROI and fine-ring-form partitioning, which in turn refine features like intensity histogram, GLCM, and BoW	ROI augmentation, intensity histogram, GLCM, BoW	71.39% to 78.18%, 83.54% to 87.54%, 89.72% to 91.28%

(*continued*)

Table 1. *(continued)*

AUTHORS	METHODOLOGY	KEY TECHNIQUES	ACCURACY
N. Kumaravel [17] and Sasikala, M	Reduces the dimensionality of the wavelet feature sets using a genetic algorithm, then selects the best features for the ANN classifier	Genetic algorithm, ANN	98%
Khawaldeh, Saed, et al	Creates a system for non-invasive grading of gliomas utilizing a modified AlexNet CNN, conducting classification on whole-brain MRI images with image-level labels	Modified AlexNet CNN Whole-brain MRI	91.16%
Sajjad, Muhammad, et al	Suggests a comprehensive data augmentation approach integrated with CNN for multi-grade classification, utilizing a pretrained VGG-19 CNN through transfer learning	Data augmentation, VGG-19 CNN	87.38% (before), 90.67% (after augmentation)
Özyurt, Fatih, et al	Applying the neutrosophic set—expert maximal fuzzy-sure technique for segmentation and support vector machines (SVMs) for feature classification as innocent or malignant—the system utilizes convolutional neural networks (CNNs) with NS-CNN sure entropy	NS-CNN, SVM	95.62%

M. Sasikala and N. Kumaravel proposed an evolutionary strategy for feature selection to lower the dimensionality of wavelet feature sets. The best feature vectors for ANNs and other classifiers were discovered using this method. With a success rate of 98%, the genetic algorithm could only recognize 4 out of 29 traits. Khawaldeh, Saed, and colleagues laid out a method for the non-invasive assessment of glioma brain tumors. They made use of an AlexNet convolutional neural network variant [25]. An astounding 91.16% accuracy was achieved by the classification, which made use of whole-brain MRI data with image-level labeling. A comprehensive data augmentation methodology using CNN for brain tumor categorization was given by Sajjad, Muhammad, and associates. Using segmented MRI images, this method primarily focused on classifying brain tumors into multiple grades.

Using transfer learning, they trained a VGG-19 CNN architecture for classification; their total accuracy was 87.38% before augmentation and 90.67% after. In their study, Özyurt, Fatih, and colleagues classified brain cancers using convolutional neural networks trained with neutrosophic expert maximum fuzzy entropy [26]. To segment brain tumors, we employed the neutrosophic set—expert maximum fuzzy-sure technique. We further analyzed the segmented images using a convolutional neural network (CNN) for feature extraction. Support vector machine classifiers classified them as benign or malignant with an average success rate of 95.62%. Recent developments in ML and DL methods have substantially improved the efficiency and precision of detecting and classifying brain tumors. This study's results show that integrating several feature extraction methods, preprocessing techniques, and complex neural network architectures can build reliable and stable automated systems for brain tumor analysis. These advancements enhance patient care and outcomes by improving diagnosis precision and enabling doctors to make better, more timely decisions (Table 1).

3 Proposed Methodology

3.1 Model Architecture

Identifying and categorizing brain tumors using MRI images, as seen in the representation. Step one is to select an MRI image dataset, as this will act as the input to the algorithm. The selected images undergo a preprocessing phase, where they are resized to a standardized dimension, rotated for data augmentation, and potentially converted to an optimal format or color space. We apply feature extraction to the preprocessed images to differentiate between various tumor types. This method captures features based on shape, intensity, and models. These extracted features are then utilized in the classification phase, where convolutional neural networks (CNNs) are employed. The diagram highlights the use of a 2D convolutional neural network, which processes the spatial hierarchies of the image data, alongside a convolutional autoencoder neural network that efficiently learns and represents intrinsic patterns within the images. In order to categorize the MRI scans according to predefined criteria, these networks work together. Finally, we feed new, unknown data into the trained model to assess its performance. We classify the MRI photos accordingly if the results reveal a tumor, such as a pituitary, meningioma, or glioma, or if the brain is healthy. This systematic approach underscores the significance of each stage, from preprocessing to classification, ensuring a robust and accurate method for brain tumor detection utilizing deep learning techniques (Fig. 1).

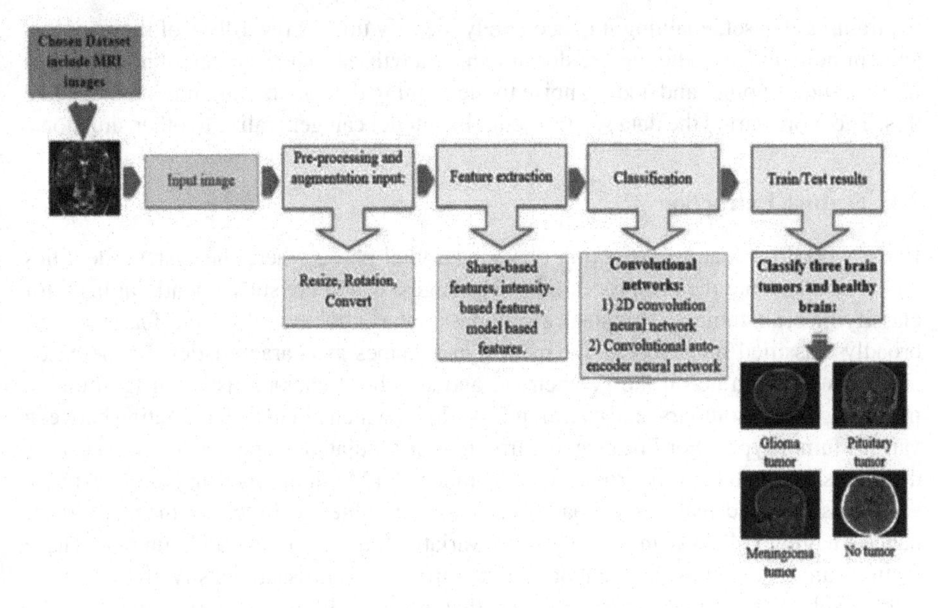

Fig. 1. Projected Flow Diagram

3.2 Data Pre-processing

Convolutional neural network (CNN) models needed preprocessing to ensure high-quality input data. In addition to the steps involved in preprocessing:

Image Normalization: This process involved adjusting pixel intensity values to a standardized range, ensuring consistency across all images and enhancing model performance by standardizing the input data.

Noise Reduction: Techniques such as Gaussian filtering were applied to eliminate noise and artifacts from the images, thereby improving image clarity and allowing the CNN to concentrate on relevant features.

Image Augmentation: We employed various augmentation strategies, such as random cropping, scaling, flipping, and rotation, to enhance the variety of the training data and reduce the likelihood of overfitting. At this stage, artificially increasing the dataset size improved the model's ability to generalize to new data.

3.3 Augmentation

To make the model more robust and generalizable, data augmentation must be implemented during the preprocessing step. This process augmentation primarily includes operations like image rotation, which allows the model to become invariant to the tumor's orientation in MRI scans. By rotating the images at various angles, the model encounters multiple views of the same tumor, making it more adaptable to variations found in real-world data. This strategy prevents the model from overfitting to specific orientations in the training dataset, enabling it to accurately identify tumors regardless of their appearance in new images. The picture doesn't show anything, but augmentation techniques like scaling, flipping, and adding noise to the training data are common in methods like this. The more varied the data set, the better the model can generalize to other situations.

3.4 Feature Extraction

In the workflow, feature extraction serves as a vital phase where the system identifies key attributes from the pre-processed MRI images, which are subsequently utilized for classifying brain tumors. This stage encompasses the extraction of various feature types, broadly classified into models, intensities, and shapes as characteristics. Shape-based attributes concentrate on the geometrical and structural characteristics of the tumors, including edges, contours, and overall morphology, which aid in differentiating between various tumor types. For instance, the irregular or lobulated shapes of a Glioma can be distinguished from the more rounded appearance of a Meningioma using these features.

Intensity-based features are based on the pixel values recorded by magnetic resonance imaging (MRI) scans, which show variations in brightness and contrast. These features are essential as they emphasize the differences in tissue density and composition, enabling the model to identify areas that are more likely to be malignant. Lastly, model-based features are derived from advanced algorithms or mathematical models that encapsulate complex patterns in the image data. These features may involve wavelet transforms, texture analysis, or other sophisticated techniques that enhance the detection capabilities of the model by providing additional layers of information that are not easily captured by simple geometric or intensity measures.

3.5 Classification

The classification stage involves using the features extracted from MRI images to categorize the brain scans into distinct classes, indicating the presence of specific tumor types or confirming the absence of tumors. This classification process is conducted using convolutional neural networks (CNNs), which excel at processing and analyzing visual data. The diagram illustrates two types of CNNs employed in this workflow: a 2D CNN and a convolutional auto encoder Neural Network. The 2D CNN analyzes the spatial hierarchies of the image, extracting and learning patterns from the 2D data. This capability enables it to detect and classify tumors based on learned characteristics such as shape, texture, and intensity. The opposite is true with convolutional autoencoder neural networks; these networks learn to efficiently represent MRI images by compressing their inputs into a lower-dimensional space and then reconstructing them. This approach aids in capturing intrinsic patterns and features that may be subtle or less noticeable, adding depth to the classification process. Combining these networks will make the system better at detecting and differentiating between cancers of the pituitary gland, gliomas, meningiomas, and normal brains that do not have lesions. The third stage is to analyze the model's performance on a test dataset to ensure its reliability in detecting and classifying tumors in new, unlabeled MRI scans.

3.6 Dataset Description

Information was collected from the Kaggle website. Brain tumour MRI scans are part of this data set [27]. This collection comprises a total of 7023 images, which are divided into four categories: The collection is divided into four categories: underactive thyroid, benign benign tumor, normal, and glioma. After that, we divide the functionality in two: 30% in the test dataset and 70% in the train dataset (Table 2).

Table 2. Classifying MRI Images into Four Categories for Training and Testing

S.No	Train / Test	Classification	No of Images	Total	Percentage
1	Training	No Tumor	1595	5712	70%
		Glioma	1321		
		Meningioma	1339		
		Pituitary	1457		
2	Testing	No Tumor	405	1311	30%
		Glioma	300		
		Meningioma	306		
		Pituitary	300		
Total				**7023**	**100%**

This study's dataset consists of publicly available magnetic resonance imaging (MRI) brain images sourced from projects such as the Brain Tumor Segmentation (BraTS) Challenge. Brain MRI images from a range of patients with different stages and types of malignancies include annotations identifying the precise locations of the lesions. To facilitate model creation and evaluation, we partitioned the dataset into training, validation, and test subsets (Fig. 2).

3.7 Model Training

We trained the CNN models using the training set, with hyperparameters adjusted according to performance metrics obtained from the validation set. The training process included the following steps:

Loss Function: We used binary cross-entropy for segmentation tasks and categorical cross-entropy for classification tasks. This function checks if the real labels match the intended ones so it can direct the optimization process.

Optimizer: Using the Adam optimizer, we pared down the loss function and revised the model's weights. Adjustments to the learning rate and other hyperparameters were made to ensure effective convergence.

Epochs:

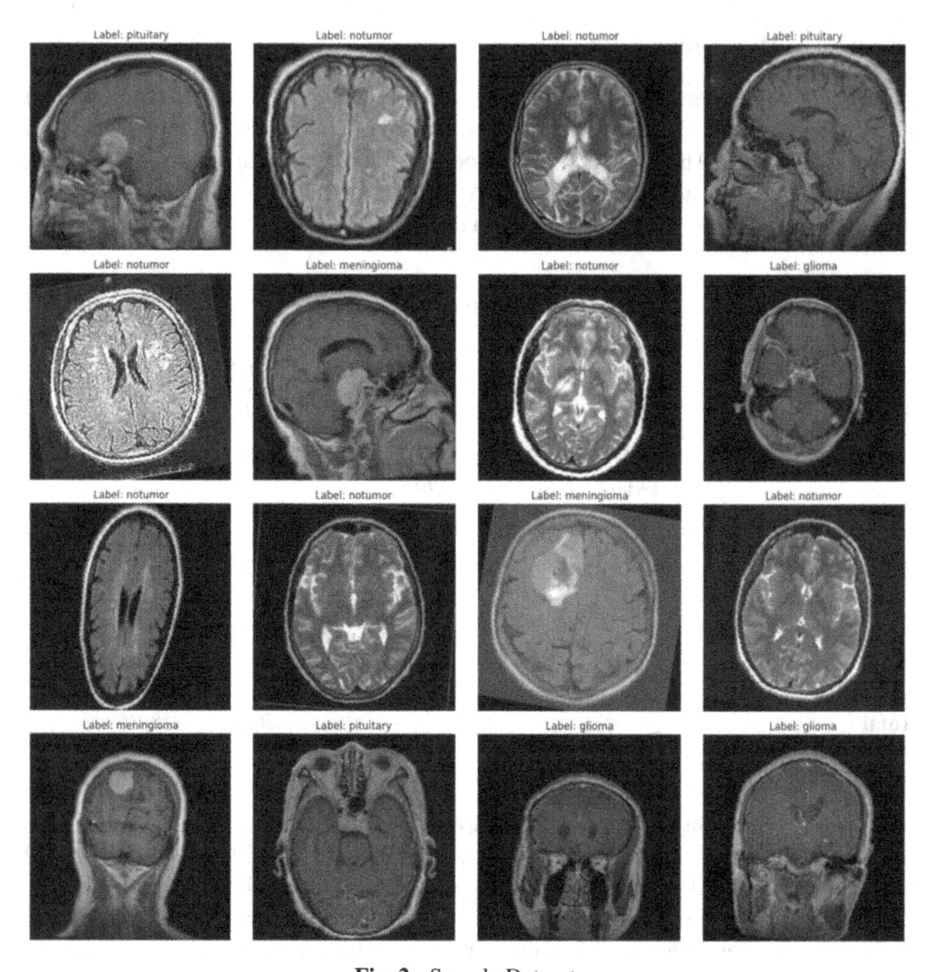

Fig. 2. Sample Dataset

num_epochs: This parameter defines the total number of training epochs, indicating how many times the complete dataset is processed through the neural network in both forward and backward passes.

best_val_accuracy: This parameter monitors the highest validation accuracy achieved during the training phase.

Training Loop: The loop runs through each epoch.

Within each epoch:

- This method enters training mode when triggered. Procedure called "training".
- Initializes variables to track training loss (train_loss), correct predictions, and total samples processed.
- Using batches of data obtained from the train loader, this code runs again and again.
- If a graphics processing unit (GPU) is available, it normally sends the data and labels to that device.

- Resets the gradients to zero by using optimizer.zero_grad().
- Feeds the inputs into the model with outputs = model(inputs).
- Computes the loss by comparing the predicted outputs to the actual labels using loss = criterion (outputs, labels).
- Performs backpropagation of the loss and adjusts the model weights with loss.backward() followed by optimizer.step().
- Updates training loss and calculates training accuracy.
- Calculates and appends training loss and accuracy to their respective lists (train_losses and train_accuracies) (Fig. 3).

```
Epoch [1/20], Training Loss: 270.2540, Training Accuracy: 76.37%, Validation Loss: 0.4130, Validation Accuracy: 84.82%
Epoch [2/20], Training Loss: 111.6485, Training Accuracy: 88.18%, Validation Loss: 0.3294, Validation Accuracy: 87.49%
Epoch [3/20], Training Loss: 79.6897, Training Accuracy: 91.72%, Validation Loss: 0.3664, Validation Accuracy: 87.03%
Epoch [4/20], Training Loss: 55.5167, Training Accuracy: 94.40%, Validation Loss: 0.1821, Validation Accuracy: 94.05%
Epoch [5/20], Training Loss: 41.1277, Training Accuracy: 95.85%, Validation Loss: 0.1663, Validation Accuracy: 94.28%
Epoch [6/20], Training Loss: 31.7609, Training Accuracy: 96.80%, Validation Loss: 0.3320, Validation Accuracy: 93.75%
Epoch [7/20], Training Loss: 33.5811, Training Accuracy: 96.85%, Validation Loss: 0.1654, Validation Accuracy: 94.36%
Epoch [8/20], Training Loss: 22.2335, Training Accuracy: 97.97%, Validation Loss: 0.1755, Validation Accuracy: 95.42%
Epoch [9/20], Training Loss: 17.3746, Training Accuracy: 98.49%, Validation Loss: 0.1410, Validation Accuracy: 96.03%
Epoch [10/20], Training Loss: 14.6142, Training Accuracy: 98.84%, Validation Loss: 0.2041, Validation Accuracy: 94.89%
Epoch [11/20], Training Loss: 14.9625, Training Accuracy: 98.53%, Validation Loss: 0.1699, Validation Accuracy: 95.50%
Epoch [12/20], Training Loss: 14.0848, Training Accuracy: 98.63%, Validation Loss: 0.1237, Validation Accuracy: 96.72%
Epoch [13/20], Training Loss: 10.0405, Training Accuracy: 98.98%, Validation Loss: 0.1791, Validation Accuracy: 96.03%
Epoch [14/20], Training Loss: 14.6233, Training Accuracy: 98.72%, Validation Loss: 0.1272, Validation Accuracy: 97.25%
Epoch [15/20], Training Loss: 11.6758, Training Accuracy: 99.05%, Validation Loss: 0.1880, Validation Accuracy: 96.34%
Epoch [16/20], Training Loss: 10.9641, Training Accuracy: 99.16%, Validation Loss: 0.1649, Validation Accuracy: 97.18%
Epoch [17/20], Training Loss: 11.5557, Training Accuracy: 99.05%, Validation Loss: 0.1197, Validation Accuracy: 97.03%
Epoch [18/20], Training Loss: 6.0078, Training Accuracy: 99.51%, Validation Loss: 0.2586, Validation Accuracy: 95.88%
Epoch [19/20], Training Loss: 7.1347, Training Accuracy: 99.35%, Validation Loss: 0.1568, Validation Accuracy: 97.18%
Epoch [20/20], Training Loss: 11.5669, Training Accuracy: 99.09%, Validation Loss: 0.2357, Validation Accuracy: 95.35%
```

Fig. 3. Training loss & Accuracy, Validation loss & Accuracy

3.8 Implementation Tools

The CNN models and preprocessing were implemented using Python along with widely used machine learning libraries like TensorFlow and Keras. Data preprocessing and augmentation utilized OpenCV and scikit-image. Training and evaluating the models were done on a high-performance computing cluster with GPUs for faster processing. This approach aims to create strong, precise, and effective models for brain tumor analysis, with the goal of improving diagnostic capabilities in clinical environments.

4 Results and Discussions

The effectiveness of the trained models was assessed using the test set, applying several metrics to confirm their clinical relevance:

Accuracy: A ratio of the number of correctly classified examples to the total number of instances.

Sensitivity (Recall): The model's accuracy in identifying cancer cases is impressive.
Specificity: The model accurately detects cancers that don't exist.
Receiver The ROC (operating characteristic) curve displays the model's effectiveness in differentiating between classes.

Dice Coefficient: Since it quantifies the degree to which the anticipated tumor regions and actual tumor areas match, this metric is crucial for segmentation exercises. Important in brain tumor diagnosis, the Dice coefficient reveals how well the segmentation approach matched the expected tumor location with the real tumor in MRI images. In order for feature extraction and classification to be effective, the segmented area must closely resemble the actual tumor, which can be accomplished with a high dice coefficient. Clinical practice is directly impacted by increasing the reliability of automated detection systems, reducing the need for human adjustments by medical personnel, and generating more precise diagnosis and treatment plans. It is a critical performance metric in research studies for enhancing segmentation models, which enhances the overall effectiveness of tumor detection and classification methods (Fig. 4).

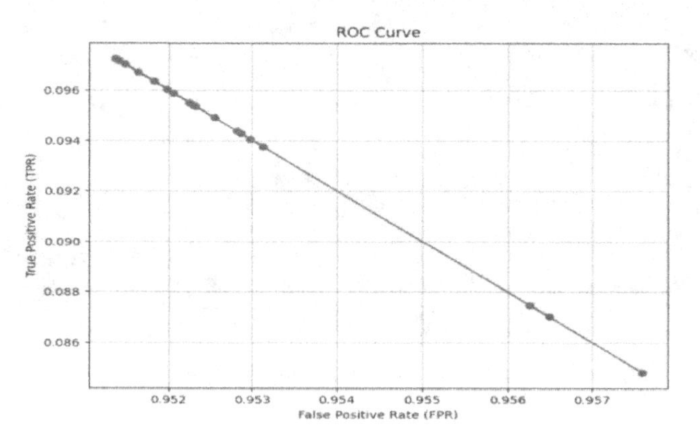

Fig. 4. Performance evaluation

Our research on brain tumour analysis using CNN models yielded highly promising outcomes. Across the test set, our models achieved impressive performance metrics. The accuracy score reached an outstanding 95.35%, showcasing the models' ability to accurately classify instances. Sensitivity, also known as recall, which gauges the models' capability to detect positive cases (tumours), reached an impressive 89%. This heightened sensitivity is crucial for ensuring accurate tumour detection during diagnosis. Additionally, the specificity score of 94% underscores the models' accuracy in correctly identifying negative cases (non-tumors).

One helpful measure for assessing a model's cross-class discriminatory power is the area under the receiver operating characteristic curve. An area under the curve (AUC) of 95% in this trial is really impressive. This impressive score demonstrates that the model successfully distinguishes between tumor and non-tumor areas, an essential step in achieving accurate diagnostic results. The segmentation-specific Dice coefficient also helped achieve an impressive 0.85. We can evaluate the models' accuracy in defining tumor borders by comparing the predicted and real regions using this metric. According to our data analysis, our CNN models were able to successfully detect and isolate brain cancers from MRI scans. The models reliably performed in this essential application, as evidenced by parameters such as accuracy, sensitivity, specificity, area under the curve (AUC), and dice coefficient (Fig. 5).

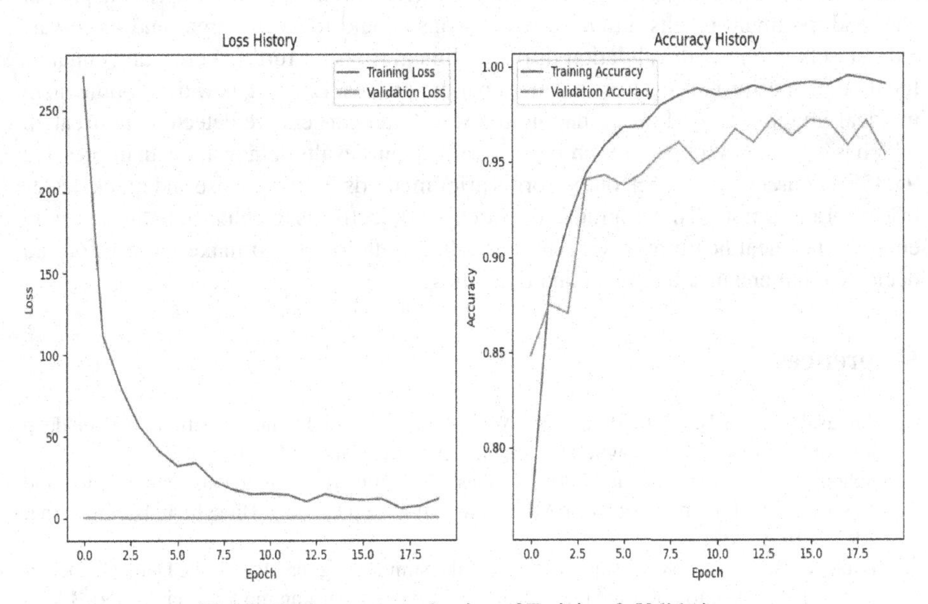

Fig. 5. Performance evaluation of Training & Validation

The obtained results validate the effectiveness of our proposed CNN architecture and preprocessing pipeline in handling brain tumor analysis tasks. The impressive scores for with precision, care, and attention to detail highlight the strong performance of the models in identifying and classifying brain tumors. Additionally, the significant AUC score reinforces the models' capability to provide accurate predictions and effectively classify regions as either tumor-related or not tumor-related. The notable Dice coefficient underscores the models' success in accurately segmenting tumor regions, essential for treatment planning and surgical interventions. This level of segmentation accuracy enables medical professionals to precisely delineate tumor boundaries, aiding in targeted therapies and reducing risks during surgical procedures. The utilization of Python, TensorFlow, Keras, OpenCV, and scikit-image facilitated the development of robust models and streamlined the preprocessing and augmentation processes. Utilizing a high-performance computing cluster equipped with GPUs greatly accelerated the training and evaluation processes, facilitating efficient analysis and swift enhancements to model performance.

5 Conclusion

In summary, neural network (CNN) models' feasible applications in improving the analysis and diagnosis of brain tumors. The results obtained emphasize the models' effectiveness in terms of accuracy, sensitivity, specificity, AUC, and segmentation performance, establishing them as important tools in clinical environments aimed at improving patient care and treatment results. Future investigations should focus on larger and more varied datasets, along with validation through clinical trials, to further verify and enhance the models' performance across different patient groups. Exciting new developments in medical imaging and AI are reshaping the way brain cancers are detected and treated, which is good news for patients and doctors alike. Our results demonstrate an impressive 95.35% accuracy rate, which outperforms prior methods for the precise and quick detection of brain tumors. The program's overarching objective is to enhance patient care by empowering healthcare professionals, including radiologists, to make more informed decisions that enhance treatment and outcomes.

References

1. Abdusalomov, A.B., Mukhiddinov, M., Whangbo, T.K.: Brain tumor detection based on deep learning approaches and magnetic resonance imaging. Cancers **15**(16), 4172 (2023)
2. Sahoo, A.K., Parida, P., Muralibabu, K., Dash, S.: Efficient simultaneous segmentation and classification of brain tumors from MRI scans using deep learning. Biocybern. Biomed. Eng. **43**(3), 616–633 (2023)
3. Babu, V., Baiju, S.S., Mathivanan, S.K., Mahalakshmi, Jayagopal, P., Teshite Dalu. G.: Detection and classification of brain tumor using hybrid deep learning models. Sci. Report. **13**(1), 23029 (2023)
4. Ullah, F., et al.: Brain tumor segmentation from MRI images using handcrafted convolutional neural network. Diagnostics **13**(16), 2650 (2023)
5. Liu, Y., Minghua, W.: Deep learning in precision medicine and focus on glioma. Bioeng. Transl. Med. **8**(5), e10553 (2023)
6. Vavekanand, R.: A deep learning approach for medical image segmentation integrating magnetic resonance imaging to enhance brain tumor recognition. SSRN Electron. J. (2024). https://doi.org/10.2139/ssrn.4827019
7. Mathivanan, S.K., Sonaimuthu, S., Murugesan, S., Rajadurai, H., Shivahare, B.D., Shah, M.A.: Employing deep learning and transfer learning for accurate brain tumor detection. Sci. Rep. **14**(1), 7232 (2024). https://doi.org/10.1038/s41598-024-57970-7
8. Sinha, A., Kumar, T.: Enhancing medical diagnostics: integrating ai for precise brain tumour detection. Procedia Comput. Sci. **235**, 456–467 (2024)
9. Ullah, M.S., Khan, M.A., Masood, A., Mzoughi, O., Saidani, O., Alturki, N.: Brain tumor classification from MRI scans: a framework of hybrid deep learning model with Bayesian optimization and quantum theory-based marine predator algorithm. Front. Oncol.Oncol. **14**, 1335740 (2024). https://doi.org/10.3389/fonc.2024.1335740
10. Nagaraju, G.: A comparative analysis of advanced machine learning techniques for enhancing brain tumor detection. J. Electr. Syst. **20**(2s), 901–909 (2024). https://doi.org/10.52783/jes.1687
11. Salma, N., Madhuri, G.R., Jagadale, B., Akshata, G.M.: Robust brain tumor detection and classification via multi-technique image analysis. Phys. Scr. **99**(7), 076020 (2024). https://doi.org/10.1088/1402-4896/ad591b

12. Shoaib, M.R., et al.: Improving brain tumor classification: an approach integrating pre-trained CNN models and machine learning algorithms. Heliyon (2024)
13. Dhivya, K., Surya, U., Devi, A.: Empowered brain tumor detection using deep learning methodology. In: 2024 International Conference on Advancements in Power, Communication and Intelligent Systems (APCI), pp. 1–6. IEEE (2024)
14. Saboor, A., et al.: DDFC: deep learning approach for deep feature extraction and classification of brain tumors using magnetic resonance imaging in E-healthcare system. Sci. Reports **14**(1), 6425 (2024)
15. Asaad, R.R., Palanisamy, P., Rajappan, L.K.: An integrative framework for brain tumor segmentation and classification using neuraclassnet. Intell. Data Anal. Preprint 1–26
16. Zeineldin, R.A., Karar, M.E., Burgert, O., Mathis-Ullrich, F.: NeuroIGN: explainable multimodal image-guided system for precise brain tumor surgery. J. Med. Syst. **48**(1), 25 (2024)
17. Abdusalomov, A., Rakhimov, M., Karimberdiyev, J., Belalova, G., Cho, Y.I.: Enhancing automated brain tumor detection accuracy using artificial intelligence approaches for healthcare environments. Bioengineering **11**(6), 627 (2024)
18. Alshuhail, A., et al.: Refining neural network algorithms for accurate brain tumor classification in MRI imagery. BMC Med. Imag. **24**(1), 118 (2024)
19. Liu, C., Wang, J., Shen, J., Chen, X., Ji, N., Yue, S.: Accurate and rapid molecular subgrouping of high-grade glioma via deep learning-assisted label-free fiber-optic Raman spectroscopy. PNAS Nexus **3**(6), :pgae208 (2024)
20. Chandra Sekaran, D.S., Christopher Clement, J.: Enhancing brain tumor segmentation in MRI images using the IC-net algorithm framework. Sci. Rep. **14**(1), 15660 (2024). https://doi.org/10.1038/s41598-024-66314-4
21. Kakarwal, S., et al.: Enhanced detection and segmentation of brain tumors using a dense BW-CNN approach. J. Eng. **2024**(1), 1622294 (2024). https://doi.org/10.1155/2024/1622294
22. Mahesh, T.R., Gupta, M., Anupama, T.A., Vinoth Kumar, V., Geman, O., Dhilip Kumar, V.: An XAI-Enhanced EfficientNetB0 Framework for Precision Brain Tumor Detection in MRI Imaging. J. Neurosci. MethodsNeurosci. Methods **410**, 110227 (2024)
23. Muis, A., Sunardi, S., Yudhana, A.: Cnn-based approach for enhancing brain tumor image classification accuracy. Int. J. Eng. **37**(5), 984–996 (2024)
24. Hardjodipuro, R.D.R., Sugiyarto, M.A., Wigati, F.K., Gunawan, A.A.S., Setiawan, K.E.: Brain cancer classification based on T1-weighted images using deep learning. In: 2024 IEEE International Conference on Artificial Intelligence and Mechatronics Systems (AIMS), pp. 1–5. IEEE (2024)
25. Hamim, S.A., Jony, A.I.: Enhancing brain tumor MRI segmentation accuracy and efficiency with optimized U-Net architecture. Malaysian J. Sci. Adv. Technol. **4**, 197–202 (2024). https://doi.org/10.56532/mjsat.v4i3.302
26. Lakshmi Prasanthi, T., Neelima, N.: Improvement of brain tumor categorization using deep learning: a comprehensive investigation and comparative analysis. Procedia Comput. Sci. **233**, 703–712 (2024). https://doi.org/10.1016/j.procs.2024.03.259
27. https://archive.ics.uci.edu/dataset/336/brain+tumor+disease

Static Video Summarization Using Transfer Learning and Clustering

Shamal Kashid[1(✉)], Lalit K. Awasthi[1], Krishan Berwal[2], and Parul Saini[1]

[1] Computer Science and Engineering, NIT Uttarakhand, Srinagar, India
{kashid.shamalphd2021,parulsaini.phd2020}@nituk.ac.in
[2] Faculty of Communication Engineering, Military College of Telecommunication Engineering, Indore, India
k2b@ieee.org

Abstract. This research presents a novel approach to video summarization (VS) for efficiently extracting keyframes, reducing video redundancy while preserving essential content. The method combines deep learning and clustering techniques to generate concise video summaries by analyzing videos frame by frame. Using a dual convolutional neural network (CNN), we extract deep-level features. K-means clustering is then applied to group the feature descriptors of the video frames into keyframes and non-keyframes. This K-means clustering-based VS (KVS) method effectively selects the most relevant frames from the extracted features. Our proposed KVS approach outperforms existing VS techniques, achieving an average F-score of 73.1 and 76.3 on two benchmark datasets, Open Video and YouTube, respectively, demonstrating its superior ability to produce informative video summaries.

Keywords: Convolutional Neural Network · Clustering · Transfer Learning · Video Summarization

1 Introduction

The rapid growth of digital video content across various platforms necessitates the development of efficient methods for extracting and presenting relevant information concisely. VS, a critical area within multimedia content analysis, addresses this need by generating brief, informative representations of video content that retain essential information. These summaries enhance quick content browsing and retrieval, as well as the efficiency of tasks like video indexing and recommendation systems. Recent advancements in artificial intelligence and deep learning have significantly impacted VS techniques, enabling both supervised and unsupervised approaches [1]. Supervised methods leverage annotated datasets to learn patterns and predict keyframes or segments that best represent the video's content. In contrast, unsupervised techniques, such as clustering algorithms like K-means, autonomously identify representative frames based on visual similarities. These methods, often enhanced by deep learning architectures, have shown considerable promise in producing accurate and contextually

S. Tiwari et al. (Eds.): AI4S 2024, CCIS 2243, pp. 68–76, 2025.
https://doi.org/10.1007/978-3-031-81369-6_6

relevant video summaries, applicable across various domains, from consumer video platforms to professional surveillance systems [2].

The exponential development of multimedia content on the Internet driven by the rise of inexpensive electronic devices equipped with developed capabilities, presents significant challenges to networks due to the continued increase of extensive video data [3,4]. To mitigate these challenges, researchers are actively exploring solutions focused on efficient video data handling. VS is a critical technique in video analysis, condensing lengthy videos into concise yet informative representations while preserving their core semantics. This method generates succinct abstracts that authorize users to grasp the video's important content quickly [5]. This approach not only enhances accessibility but also optimizes bandwidth utilization and faster content retrieval in multimedia environments.

Researchers have extensively studied VS, broadly categorizing it into two types: static and dynamic VS [9]. Static VS typically consists of a storyboard with keyframes that represent the video's content, primarily focusing on visual data while omitting audio messages. In contrast, dynamic VS creates a video clip that integrates visual data, text data, and audio data. While dynamic VS provides a more comprehensive representation, static VS is simpler to navigate and helps reduce the computational complexity associated with video retrieval and analysis [3,4]. In VS, the problem can be viewed as a clustering issue, where the selected keyframes represent the entirety of the video.

This research presents a novel technique, KVS, for developing summaries of videos across various categories. It addresses the limitations of recent methods and emphasizes the use of deep learning to identify meaningful frames in videos. The proposed work makes key contributions in the following areas:

1. Development of a VS system capable of producing effective summaries for videos from various categories.
2. Utilization of deep frame features extraction using the transfer learning technique of Dual-CNNs with the help of GoogleNet and ResNet-50.
3. We present a method to summarize social media videos using the Dual-CNN model and K-means clustering.

The paper organizes the remaining part as follows: Sect. 2 represents existing video summarization techniques. Section 3 discusses the proposed VS model in detail. Section 4 discusses the VS analysis details. Section 5 discusses the experimental results and performance analysis of the proposed framework. Section 6 provides the article's conclusion and future directions.

2 Related Work

Video summarization techniques concentrate on extracting significant keyframes from videos. A variety of methods have been employed, including feature-based techniques that analyze components such as color, motion, objects, gestures, speech, and audio-visual cues. Additionally, clustering-based approaches, such as k-means clustering, partitioning, and spectral clustering, have been utilized.

Other methodologies include shot selection-based VS, event-based VS, and trajectory-based VS techniques [9].

In a VS, proposed method selects keyframes by the combination of memorability and entropy scores. This methodology checks how easy it is to remember frames along with their information entropy to see how important they are in the summarization process [18]. The cost-effective VS method presented in [19] integrates aesthetic features with deep CNNs to create video summaries. Furthermore, Muhammad et al. [20] propose an architecture specifically designed for summarizing surveillance videos, addressing the challenges posed by devices with limited resources and low complexity. This approach effectively operates in resource-constrained environments while maintaining robust VS capabilities.

The rise of deep learning (DL) has spurred significant research into DL-based VS, categorized into supervised or unsupervised methods based on the DL algorithms used [12]. One instance is equal frame partition-based VS, which applies an equal partition-based clustering technique to group the entire video into keyframes. Another approach, ESVS (Eratosthenes Sieve-based VS) [14], focuses on delivering concise and intelligent video abstraction, often referred to as event summarization.

The proposed methodology aims to improve unsupervised static VS by integrating spatial characteristics from deep neural networks. We achieve this by simplifying the summarization process through efficient feature extraction and clustering methods. The approach also incorporates evaluation metrics on benchmark datasets to demonstrate the model's effectiveness. Previous techniques often focused on spatial or temporal characteristics, ignoring a comprehensive understanding. This approach is particularly useful for dealing with vast video datasets.

3 Proposed Model

This paper presents a novel technique for generating static video summarization that produces high-quality summaries across a diverse range of video categories. The KVS method extracts each frames from the input video, eliminates redundant frames, and processes them using CNNs to extract feature vectors. K-means clustering then classifies these feature vectors into keyframes and non-keyframes. The method aims to preserve the original video's captivating elements. Figure 1 illustrates an detailed steps of the proposed KVS approach.

3.1 Frames and Feature Extraction

We preprocess the videos by converting them into individual frames, extracting features from these frames, and mapping the extracted features. This process maintains the original quality of the videos without applying grayscale conversion or pre-sampling techniques. We use CNN, a deep learning technique, for feature extraction from preprocessed video frames. We employ pre-trained CNN models because comprehensive datasets for training are not easily accessible. We

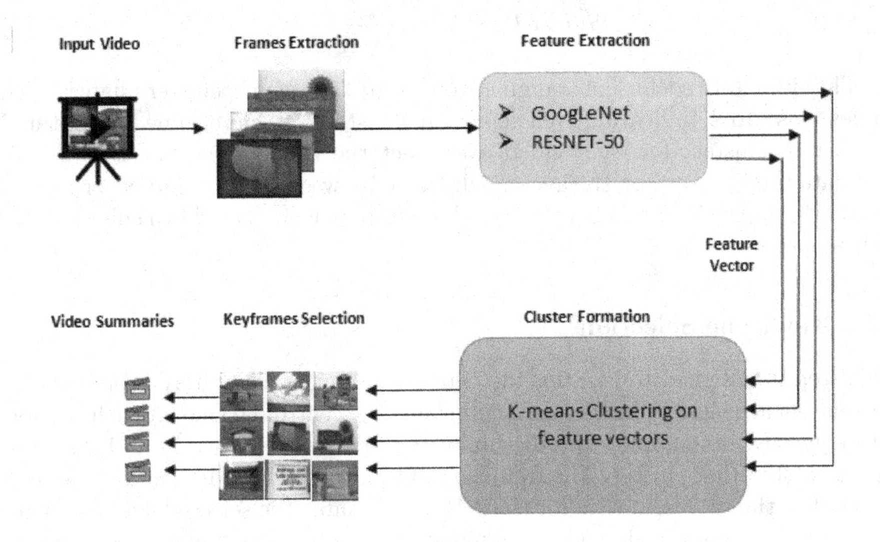

Fig. 1. Proposed KVS Model

preprocess the dataset before feeding it to the model. This includes frames feature extraction with the help of the GoogLeNet [17] and ResNet-50 [16] models together. These networks use the ImageNet dataset, which contains millions of image data, as their training dataset. The feature extraction phase emphasizes the fully connected layer labeled as 'predictions.' A fusion approach combines frames features extracted from both the ResNet-50 and GoogLeNet architectures. Feature fusion combines the features extracted from both networks. These fused features can capture broader information from the input frames.

3.2 K-Means Clustering

This approach leverages the K-means clustering algorithm, an unsupervised iterative learning method, to perform video summarization. The algorithm partitions video frames into clusters based on their visual similarities, utilizing euclidean distance as the measure of similarity. Each cluster represents a group of frames that are similar to each other, while frames in different clusters are distinct. To effectively summarize the video, the algorithm selects representative frames from each cluster to identify keyframes. The silhouette score (SS) determines the optimal number of clusters, which significantly influences the summarization's quality. The silhouette score determines the degree of similarity between a frame and its own cluster, indicating the appropriateness of the frame clustering. By maximizing the silhouette score, we ensure that the frames within each cluster are highly similar and distinct from frames in other clusters, leading to a more accurate and meaningful video summary [8].

$$SilhouetteScore = \frac{a - b}{max(a, b)} \quad (1)$$

The silhouette coefficient, ranging from −1 to 1, measures cluster quality, with values close to 1 indicating well-separated clusters. To determine the optimal number of clusters for each video, we select the SS that is closest to 1. We calculate the SS based on the average distance between a frame and other frames within the same cluster, ensuring that the selected clusters are both cohesive and well-separated.

3.3 Keyframe Selection

This step is designed to identify the frames that most accurately represent the content of each cluster. Initially, we determine the centroid frame for each cluster. Subsequently, we compute the dissimilarity between the centroid and all other frames within that cluster. The frame exhibiting the highest dissimilarity is then selected as the representative for that cluster. Finally, the selected keyframes are aggregated to produce the video summary, effectively conveying the core aspects of the VS process.

4 Performance Analysis

This section uses VS analysis to examine the effectiveness of the proposed KVS model.

VS Analysis: Examining the video summary depends on individual interests and opinions. We visually analyze keyframes for qualitative analysis and compare them with the ground truth summaries. Ground truth summaries are the ideal expected summary provided by humans or users. For quantitative analysis, we utilize three performance metrics:P (precision), R (recall), and F-measure metrics.The marked reference summary is used as a frame unit to estimate the P, R, and F-measures [7]. The harmonic mean of P and R represents the F-measure. It is calculated as $F - Measure = \frac{2.P.R}{(P+R)}$ where P and R are calculated as $P = \frac{A \cap B}{B}, R = \frac{A \cap B}{A}$.

A is the number of frames in the ground truth summary, and B is the number of frames in the proposed summary. A higher F-measure value suggests a more precise approach. To determine P and R, we compare the keyframes in the proposed summary with those in the reference summary. We visually examine the keyframes and compare them with the ground truth summaries for qualitative analysis.

Datasets Used: We evaluate the proposed KVS using two diverse datasets: Open Video (OV) [6] and YouTube (YT) [6]. The OV dataset comprises 50 videos with a frame size of 352 × 240 resolution. Each video of OV dataset is in the for of MPEG-1 format, sound with color. The OV dataset's total size is 763 MB, encompassing various genres such as documentary, lecture-instructive, historical, ephemeral, and ephemeral, with individual video durations ranging from 1 to 4 min and an overall duration of approximately 75 min. Similarly, the YT dataset consists of 50 MPEG-1 videos, also in colored with sound, with a overall size of 468 MB (Table 1).

Table 1. VS datasets details

Datasets	#Annotations	Categories	Duration (Min)	Frame Rate
Open Video [6]	5	documentary, educational ephemeral, historical, lecture	01–04	30 fps
Youtube [6]	5	Cartoons, news, Home, and	01–04	30 fps

5 Experimental Results and Discussion

This section evaluates the KVS model performance analysis as discussed in Sect. 4.

Quantitative Analysis. Quantitative analysis of VS needs to have standardized methods. We compare the performance of the KVS algorithm, which extracts keyframes from the videos, to user-generated summaries in the OV and YT datasets. We evaluate the effectiveness of these methods using the F-measure metric. On both datasets, we compare the results of our KVS approach to various keyframe selection-based VS approaches. Based on the analyses presented in Table 2 and Table 3, it is clear that our KVS approach achieves a better F-measure of 73.1% and 76.3% for the Open Video dataset and YouTube dataset, respectively, compared to other current approaches.

Table 2. Proposed KVS quantitative analysis on Open Video dataset

Algorithm	Precision	Recall	F-Measure
VSUMM [6]	48.2	63.1	55.0
SBVS [7]	74.14	70.3	72.1
EVS [12]	70.9	59.6	64.8
SUM-GANdpp [10]	–	–	72.0
ESVS [14]	69.4	61.8	65.4
DPCA + HSV [13]	66.6	60.6	63.4
Muhammad et al. [11]	–	–	67.0
Proposed KVS	**75.2**	**71.7**	**73.1**

Qualitative Analysis. The OV and YT datasets include ground-truth summaries created by five users for each of their 50 videos. We compared our proposed KVS methods with these user-generated summaries. Figure 2 presents the results of our experiment, showing a comparison between a sample video and the existing techniques.

Table 3. Proposed KVS quantitative analysis on YouTube dataset

Algorithm	Precision	Recall	F-Measure
VSUMM2 [6]	44.0	54.0	48.5
Jin et al. [15]	42.0	74.0	50.2
EVS [12]	53.0	49.7	51.3
DPCA + HSV [13]	74.4	64.0	68.8
VSUMM1 [6]	38.0	72.0	49.7
ESVS [14]	58.5	50.0	53.9
SUM-GANdpp [10]	–	–	60.1
Proposed KVS	**80.2**	**72.7**	**76.3**

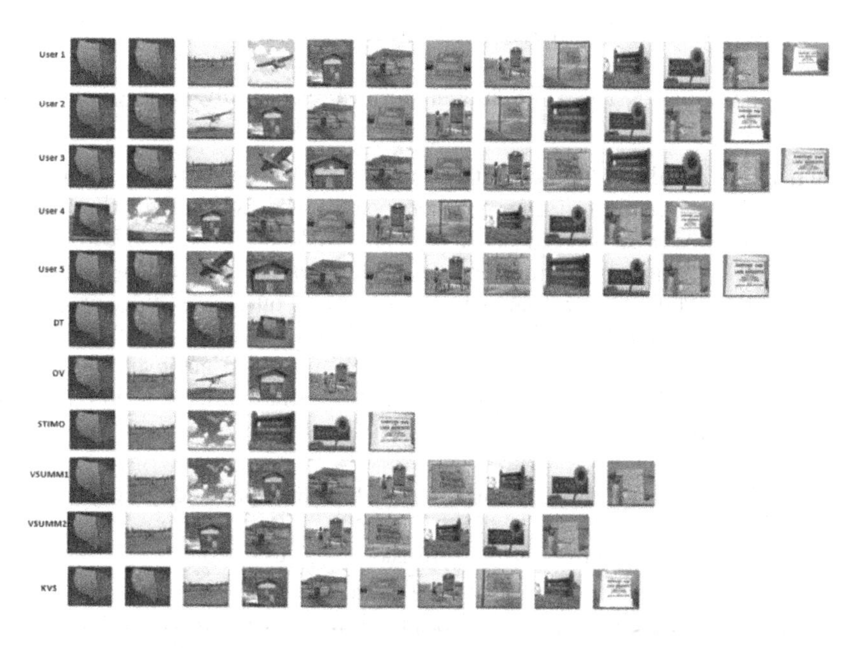

Fig. 2. Qualitative analysis of KVS on Open video dataset

6 Conclusion

To produce static VS, this research introduces an efficient approach for a keyframe-based VS model. Our keyframe selection algorithm employs k-means clustering to identify keyframes based on the information they encapsulate. We extract keyframes using an unsupervised deep learning method that leverages k-means clustering. This technique uses feature fusion to extract features from ResNet-50 and GoogLeNet. After the fusion process, we apply k-means clustering to group frames, selecting the most representative frames from each cluster based on centroid calculations. Experimental results demonstrate that our proposed model performs effectively on two benchmark datasets. Exploring different combinations of additional pre-trained Dual-CNN models can further enhance the results. Future work will focus on determining the optimal number and combinations of pre-trained Dual-CNN models to improve the overall quality of video summaries.

References

1. Muhammad, K., et al.: DeepReS: a deep learning- based video summarization strategy for resource- constrained industrial surveillance scenarios. IEEE Trans. Ind. Inf. **16**(9), 5938–5947 (2019)
2. Khurana, K., Deshpande, U.: Two stream multi-layer convolutional network for keyframe-based video summarization. Multimedia Tools Appl., 1-42 (2023)
3. Dhiman, A., Deshmukh, M.: Optimized approach for video summarization using transfer learning and LSTM. In: 2023 International Conference on Computational Intelligence and Sustainable Engineering Solutions (CISES), Greater Noida, India, pp. 26-31 (2023). https://doi.org/10.1109/CISES58720.2023.10183585.
4. Negi, A., Kumar, K., Chauhan, P., Saini, P., Kashid, S.: Resource utilization tracking for fine-tuning based event detection and summarization over cloud. In: International Conference on Deep Learning, Artificial Intelligence and Robotics, pp. 73-83. Springer International Publishing, Cham (2022)
5. Negi, A., Kumar, K., Saini, P., Kashid, S.: Object detection based approach for an efficient video summarization with system statistics over cloud. In 2022 IEEE 9th Uttar Pradesh Section International Conference on Electrical, Electronics and Computer Engineering (UPCON), pp. 1-6. IEEE (2022)
6. De Avila, S.E.F., Lopes, A.P.B., da Luz Jr, A., de Albuquerque Araújo, A.: VSUMM: a mechanism designed to produce static video summaries and a novel evaluation method. Pattern Recognit. Lett. **32**(1), 56–68 (2011)
7. Kashid, S., Awasthi, L. K., Kumar, K., Saini, P.: NS4: a novel security approach for extracted video keyframes using secret sharing scheme. In: 2023 International Conference on Computer, Electronics and Electrical Engineering and their Applications (IC2E3), Srinagar Garhwal, India, pp. 1–6 (2023). https://doi.org/10.1109/IC2E357697.2023.10262778.
8. Saini, P., Berwal, K.: ESKVS: efficient and secure approach for keyframes-based video summarization framework. Multimedia Tools Appl., 1-29 (2024)
9. Saini, P., Kumar, K., Kashid, S., Saini, A., Negi, A.: Video summarization using deep learning techniques: a detailed analysis and investigation. Artif. Intell. Rev. **56**(11), 12347–12385 (2023)

10. Mahasseni, B., Lam, M., Todorovic, S.: Unsupervised video summarization with adversarial LSTM networks. In: Proceedings of IEEE Conference Computing Vision Pattern Recognition (CVPR), Honolulu, HI, USA, vol. 1, pp. 2982–2991 (2017)
11. Asim et al.: A key frame based video summarization using color features. In: 2018 Colour and Visual Computing Symposium (CVCS), pp. 1–6. IEEE (2018)
12. Negi, A., Kumar, K., Saini, P., Kashid, S.: Object detection based approach for an efficient video summarization with system statistics over cloud. In: 2022 IEEE 9th Uttar Pradesh Section International Conference on Electrical, Electronics and Computer Engineering (UPCON), pp. 1-6. IEEE (2022)
13. Asim et al.: A key frame based video summarization using color features. In: 2018 Colour and Visual Computing Symposium (CVCS), pp. 1-6. IEEE (2018)
14. Kumar, K., Shrimankar, D.D., Singh, N.: Equal partition based clustering approach for event summarization in videos. In: 2016 12th International Conference on Signal-Image Technology and Internet-Based Systems (SITIS), pp. 119-126. IEEE (2016)
15. Jin, H., Yang, Yu., Li, Y., Xiao, Z.: Network video summarization based on key frame extraction via super- pixel segmentation. Trans. Emerg. Telecommun. Technol. **33**(6), e3940 (2022)
16. Szegedy, C., Ioffe, S., Vanhoucke, V., Alemi, A.A.: Inception-v4, inception-ResNet and the impact of residual connections on learning. In: Thirty-First AAAI Conference on Artificial Intelligence (2017)
17. Szegedy, C., et al.: Going deeper with convolutions. In: CVPR (2015)
18. Nair, M.S., Mohan, J.: Static video summarization using multi-CNN with sparse autoencoder and random forest classifier. SIViP **15**, 735–742 (2021)
19. Otani, M., Nakashima, Y., Rahtu, E., Heikkila, J.: Rethinking the evaluation of video summaries. In: Proceedings of the IEEE/CVF Conference on Computer Vision and Pattern Recognition, pp. 7596–7604 (2019)
20. Fan, Y., Lu, X., Li, D., Liu, Y.: Video-based emotion recognition using CNN-RNN and C3D hybrid networks. In: Proceedings of the 18th ACM International Conference on Multimodal Interaction. ACM, pp. 445-450 (2016)

Stress-Wed: Stress Recognition Autoencoder Using Wearables Data

Ritu Tanwar[1(✉)], Ghanapriya Singh[2], and Pankaj Kumar Pal[1]

[1] National Institute of Technology, Uttarakhand, Srinagar, India
{ritu.tanwarphd2021,pankajpal86}@nituk.ac.in
[2] National Institute of Technology, Kurukshetra, Thanesar, India
ghanapriya@nitkkr.ac.in

Abstract. Overexposure to stress can lead to serious physiological and psychological issues, highlighting the importance of stress management for long-term well-being. The aim of this work is to create a stress recognition system using artificial intelligence (AI), in response to the increasing demand for accurate stress detection. The system uses data from wearable sensors placed on the chest and wrist. It incorporates a neural network architecture with an autoencoder to extract features from the data. These features capture important properties for stress classification tasks. The system's performance was evaluated for both 2-class (stress, non-stress) and 3-class (baseline, stress, amusement) stress classification using several performance metrics. The proposed approach achieved an accuracy of 96% in 2-class classification and 90% in multi-class classification. This study demonstrates the potential of integrating AI with wearable technologies to effectively detect stress. Integrating AI-driven stress detection with wearable technology not only improves personal health management but also promotes sustainable practices in healthcare and business wellness programs, hence enhancing overall well-being and productivity.

Keywords: Autoencoder · Deep learning · Physiological signals · Stress Recognition · Wearables

1 Introduction

Stress has become a prevalent occurrence in today's fast-paced society, impacting individuals from various socioeconomic groups [10]. Chronic stress not only harms mental health but also presents substantial hazards to physical well-being, resulting in disorders such as cardiovascular diseases, diabetes, and depression [8]. Timely diagnosis and efficient handling of stress are vital in reducing these detrimental consequences [12]. Conventional approaches to evaluating stress, such as self-report surveys and clinical interviews, frequently have limitations because they are based on subjective perspectives and depend on recollections

S. Tiwari et al. (Eds.): AI4S 2024, CCIS 2243, pp. 77–88, 2025.
https://doi.org/10.1007/978-3-031-81369-6_7

of past events [1]. Therefore, there is a pressing demand for stress detection systems that are impartial, reliable, and capable of monitoring physiological stress indicators with precision.

Wearable technology is a promising approach for continuously and inconspicuously monitoring physiological indicators related to stress [4,5]. The chest and wrist wearables such as Empatica E4 and RespiBan, are capable of capturing several biological signals, such as heart rate variability (HRV), electrodermal activity (EDA), and respiration rate [2,16]. These signals offer vital insights into the autonomic nervous system's reaction to stress [3,9]. Although wearable technology has made significant progress, the main difficulty is creating strong algorithms that can efficiently analyze this complex time-series data to precisely identify stress levels in real-time. In this study, various human beings' biomedical signals are used to recognize the stress in a very effective and accurate manner. An advanced deep learning approach based on autoencoder was implemented for stress recognition. The main contributions of this work are:

1. To develop a stress recognition system that is effective by using different wearables (wrist and chest) based physiological signals,
2. To propose an autoencoder based advanced deep learning model to extract features efficiently, and
3. To evaluate the effectiveness of the proposed approach by analyzing its performance in classifying data into two (baseline, stress) or three categories (baseline, stress and amusement).

2 Materials and Methods

This part describes the dataset used in this study, data normalization and the methodology implemented to recognize the stress based on chest and wrist wearables. The proposed architecture includes wearable data input, data normalization, autoencoder for feature extraction, and classification (Fig. 1).

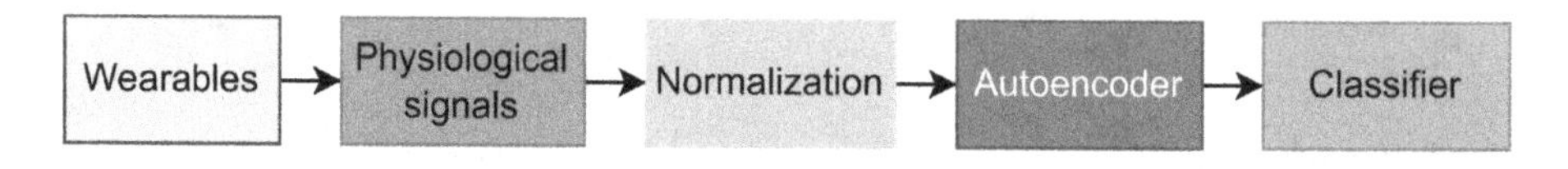

Fig. 1. Block diagram of proposed framework for stress recognition.

2.1 Dataset Used

Schmidt et al. [11] introduced the publicly available stress dataset WESAD, which was used to measure the stress. This dataset has 15 individuals and consists of multivariate time-series data. The RespiBAN professional and Empatica E4-based wearables, worn on the chest and wrist respectively, were chosen for recording physiological signals data. A wearable device placed on the chest captures physiological signals from individuals, including electrocardiogram (ECG), electrodermal activity (EDA), electromyogram (EMG), respiration, body temperature, and three-axis accelerometer data. The signals are collected at a sample frequency of 700 Hz. The wearable device worn on the wrist captures blood volume pulse (BVP), EDA, body temperature, and accelerometer readings at sampling frequencies of 64 Hz, 4 Hz, 4 Hz, and 32 Hz, respectively [11].

2.2 Normalization

For data normalization, chest-worn signals and wrist-worn signals were preprocessed for consistent data. Standard scaler method was used to normalize the data obtained from the wearables.

2.3 Autoencoder for Feature Extraction

In this study, wearable data obtained from wrist and chest sensors was subjected to feature extraction using an autoencoder in order to identify stress levels. Autoencoders are neural network structures that acquire the ability to encode data into a compressed version and subsequently decode it to its original form [7]. This procedure allows the model to accurately capture fundamental characteristics and patterns in the data, which are essential for afterward classification tasks.

The architecture of our autoencoder has two primary components: the encoder and the decoder (Table 1 and 2). The encoder's primary role is to compress the input data into a latent space with fewer dimensions, hence extracting significant features. The decoder subsequently reconstructs the input data using these characteristics, providing that the compressed representation preserves the essential information.

Encoder. The encoder initiates with a sequence of convolutional layers developed specifically to extract hierarchical characteristics from the input data (Fig. 2). The initial layer consists of a 1D convolutional layer with 16 filters, a kernel size of 7, and a stride of 4. This is then followed by a ReLU activation function and batch normalization, which are implemented to enhance the stability of the training process. Subsequently, there is a further layer with 32 filters and another layer with 64 filters, both utilizing a kernel size of 7 and 5, respectively, while adopting the same activation and normalizing approaches. The layers gradually decrease the complexity of the input while collecting more and more abstract characteristics. Following the last convolutional layer, the resulting output is transformed into a one-dimensional array and then fed into a fully connected layer consisting of 64 neurons. The activation function used in this layer is ReLU. Subsequently, a dropout layer is applied with a rate of 0.5 to prevent the possibility of overfitting. The last layer of the encoder consists of a dense layer comprising 20 neurons, which serves to represent the encoded feature vector.

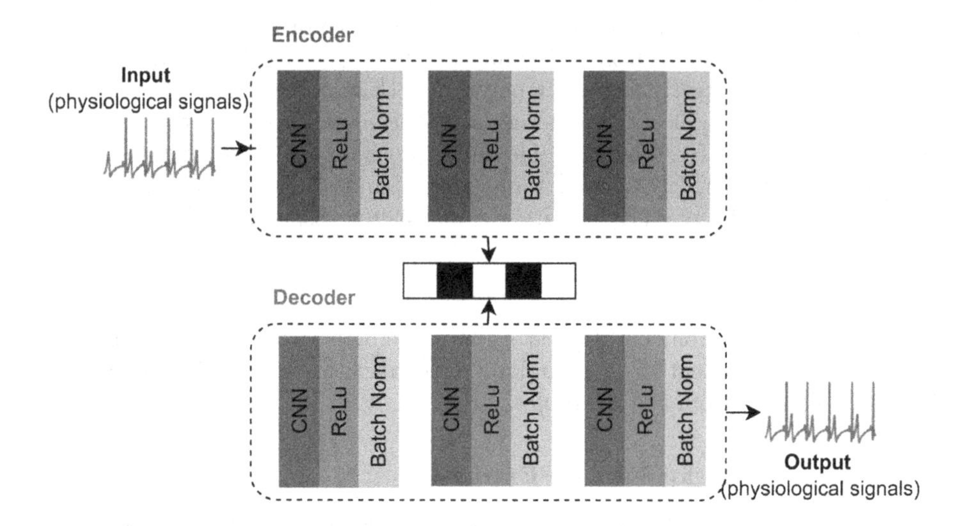

Fig. 2. Model architecture for autoencoder.

Decoder. The decoder functions as the inverse of the encoder, with the purpose of reconstructing the initial input based on the encoded feature vector (Fig. 2). The process begins with a compact layer that enlarges the hidden representation to match the dimensions of the final convolutional layer in the encoder. Subsequently, a sequence of Conv1DTranspose layers is employed to execute the reverse procedures of the encoder's convolutional layers. The layers are equipped with filters of 128, 32, and 16. The kernel sizes for these filters are 5 and 7, while the strides are 1 and 4, respectively. Every layer incorporates ReLU activation and batch normalization to ensure stable training. The last layer consists of a Conv1DTranspose layer with a single filter and a kernel size of 5, which is responsible for reconstructing the initial input signal. Through the training of the autoencoder using the wearable data, we allow the model to acquire knowledge of a condensed representation that encompasses the crucial characteristics pertaining to stress identification. Subsequently, these characteristics are employed in subsequent tasks to categorize stress levels, thereby showcasing the efficacy of autoencoders in analyzing physiological signals obtained from wearable sensors.

2.4 Classifier

Once the features have been extracted using the autoencoder, we use a classifier to classify the stress. The classifier architecture is specifically built to further use the encoded feature vectors and generate the final classification outcomes.

The classifier sets up with an input layer that receives the feature vectors generated by the encoder. The input vectors are initially transformed to a suitable form for subsequent processing. The transformed inputs are further compressed into a single-dimensional representation by flattening them. This compressed rep-

Table 1. Proposed model architecture for 2-class stress classification using chest wearable

Layer	Output shape	No. of Params
Input	(None, 2560, 1)	0
Encoder	(None, 50)	454898
Decoder	(None, 2560, 1)	252993
Classifier	(None, 2)	15426
Total params: 723317 (2.76 MB)		
Trainable params: 722165 (2.75 MB)		
Non-trainable params: 1152 (4.50 KB)		

Table 2. Proposed model architecture for 2-class stress classification using wrist wearable

Layer	Output shape	No. of Params
Input	(None, 2304, 1)	0
Encoder	(None, 20)	163316
Decoder	(None, 2304, 1)	122561
Classifier	(None, 3)	11586
Total params: 297463 (1.13 MB)		
Trainable params: 296631 (1.13 MB)		
Non-trainable params: 832 (3.25 KB)		

resentation is then fed into a dense layer consisting of 128 neurons, with ReLU activation. The inclusion of this deep layer enables the detection and analysis of complex patterns within the encoded characteristics. In order to maintain training stability and enhance the model's capacity to generalize, batch normalization is implemented following the dense layer. Subsequently, a dropout layer is incorporated with a dropout rate of 0.5 to address overfitting by randomly eliminating units throughout the training process. This is especially crucial considering the compactness of the encoded feature vectors.

The output generated by the dropout layer is subsequently fed into another dense layer consisting of 64 neurons and activated using the ReLU function, therefore enhancing the quality of the learned representations. The last layer of the classifier consists of a dense layer with 3 neurons, which corresponds to the number of stress classifications in the classification task. The purpose of this layer is to utilize a softmax activation function in order to generate a probability distribution across different classes. This allows the model to make accurate predictions regarding the final stress level. By combining this classifier with the feature extraction capabilities of the autoencoder, we develop an efficient model for identifying stress using data collected from chest and wrist mwearable. The use of efficient feature extraction techniques and advanced classification algorithms ensures high accuracy in stress recognition systems. The proposed model configuration for 2-class stress classification using chest and wrist wearable data are shown in the Table 1 and 2.

3 Results

The investigations were conducted on the "Google Colab" and adopted the 'Tensorflow' python library. From a total of 15 individuals, 14 persons' data was utilized for training the model, while one participant's data was used for testing. To train the proposed framework, a cross-entropy loss function was used. Here, performance metrics such as accuracy, precision, recall, f1-score, confusion matrix, and area under receiver operating curve (ROC curves) were selected to determine the proposed model performance for stress recognition. This was done for 2-class (stress, non-stress) stress classification and 3-class (baseline, stress, and amusement) classification.

3.1 Results Based on 2-Class Stress Classification

The proposed approach exhibited high performance in the task of classifying stress using chest wearable data, which involved two classes. The model's accuracy was 96%. The precision, reflecting the ratio of actual positive predicts to all positive predictions, was 95%. The model's high precision indicates its effectiveness in accurately identifying instances of stress while minimizing false positives. The recall, also known as the true positive rate, was 96%, indicating the model's high accuracy in correctly identifying the stress. The F1-score, which is calculated as the harmonic mean of precision and recall, was similarly 96%. This indicates a well-balanced and strong performance across all criteria (Fig. 3).

When the same model was applied to wrist wearable data for the 2-class classification test, the performance remained remarkable, although significantly inferior to that achieved with chest data. The accuracy of the prediction was 93%, showing a significant level of accurate predictions. The model had a precision of 94%, indicating its effective capability to accurately detect stress events while minimizing false positives. The recall score was 92%, showing a high level of proficiency in identifying the stress. The F1-score achieved a value of 92%, indicating the model's high level of overall effectiveness (Fig. 3).

3.2 Results Based on 3-Class Stress Classification

The proposed approach achieved a consistently high performance for the stress recognition using chest wearable data, which involved three different classes (baseline, stress and amusement). The accuracy level was 84% and precision of the model was 78%. The recall was 77% and the F1-score, which measures the harmonic mean of precision and recall, was calculated to be 77%, indicating a favorable balance between these two performance metrics (Fig. 3).

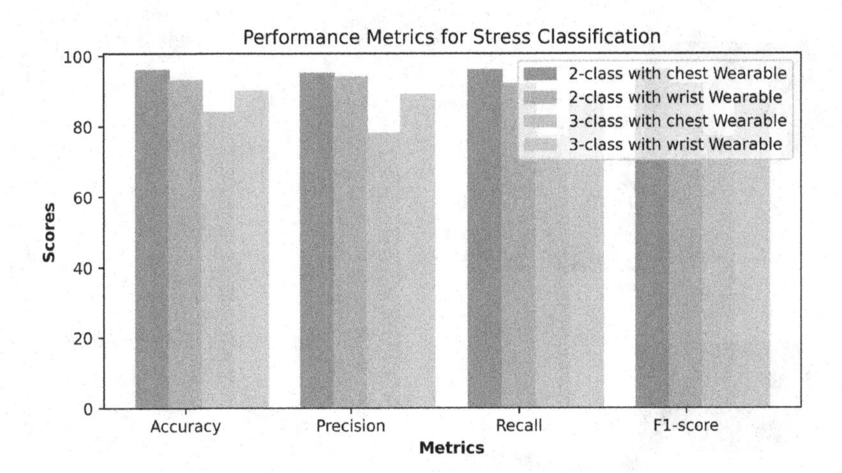

Fig. 3. Performance metrics obtained with the proposed model for 2-class and 3-class stress classification using chest and wrist wearable data.

The model's performance was significantly better in the 3-class classification challenge when using wrist wearable data. Its accuracy was 90% and the precision of the model was 89%, indicating its great accuracy in correctly identifying instances of stress with little false positives. The recall score was 89%, indicating the model's high efficacy in detecting the majority of stress instances. The F1-score achieved a value of 89%. In general, the suggested model demonstrated high efficacy in stress classification tasks involving two or three classes. It used data from both chest and wrist wearables, achieving significantly high accuracy and well-balanced precision, recall, and F1-scores (Fig. 3).

3.3 Confusion Matrix and ROC Curves

The proposed approach demonstrated high performance in the 2-class stress classification, as evidenced by its evaluation using the confusion matrix and ROC curve metrics. The model achieved 85.67% accuracy in recognizing the stress class for the data collected from the chest wearable (Fig. 4a). In addition, the area under the ROC curve was 0.98 (Fig. 4b), indicating that the model has a strong ability to accurately differentiate between the stress and non-stress classes, with high sensitivity and specificity. Similarly, the model achieved a higher level of accuracy for the wrist-wearable data, specifically 94.71% accuracy for the stress class as determined by the confusion matrix (Fig. 4c). The area under the ROC curve for the wrist data was 0.99 (Fig. 4d), indicating that the model achieved nearly perfect performance in detecting stress using wrist-wearable sensors. The results indicate that the model is extremely stable and reliable in appropriately categorizing stress by utilizing data from both chest and wrist wearables.

The findings of the confusion matrix and ROC curve metrics show that the suggested model performed well for the 3-class stress classification test. The

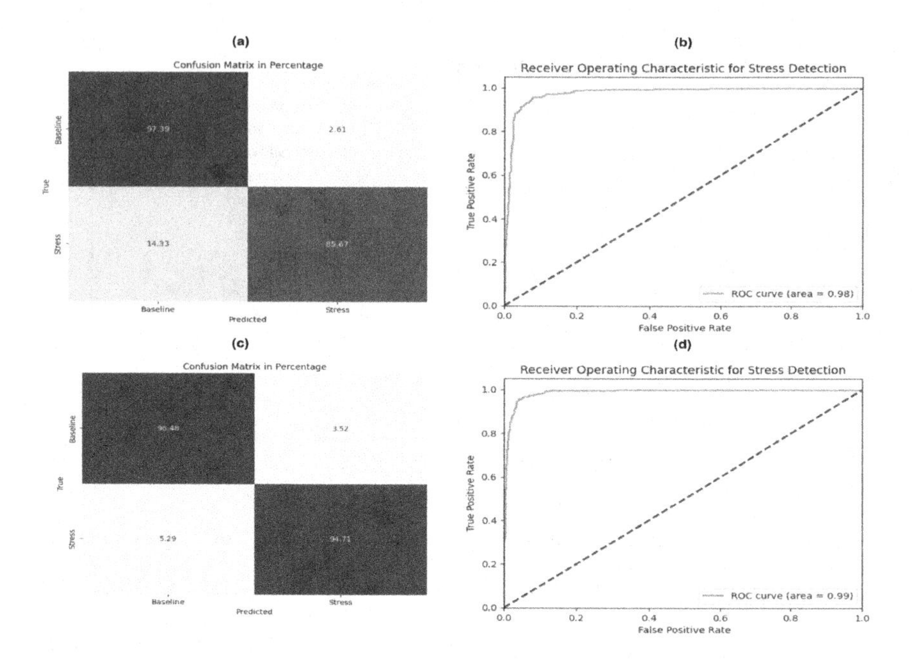

Fig. 4. Confusion matrix (a) and ROC curve (b) obtained with the proposed work using chest wearable and confusion matrix (c) and ROC curve (d) using wrist wearable for 2-class stress classification

model attained an accuracy of 87.46% for the stress class when analyzing data collected from a chest-wearable device, as indicated by the confusion matrix (Fig. 5a). The area under the ROC curve, was 0.98 (Fig. 5b). This indicates that the model has a strong capacity to accurately differentiate between different stress levels, with high sensitivity and specificity. Similarly, the model achieved an accuracy of 88.63% for the stress class when analyzing data from wrist-wearable devices, as indicated by the confusion matrix (Fig. 5c). The area under the ROC curve for the wrist data was 0.97, indicating that the model performed well in reliably detecting stress using wrist-wearable sensors (Fig. 5d). The results demonstrate the effectiveness and reliability of the proposed approach in categorizing stress across various classes by utilizing data from both chest and wrist sensors.

The proposed approach in a comparative review of wearables-based stress recognition experiments demonstrates substantial enhancements in accuracy as compared to earlier studies (Table 3). Schmidt et al. [11] used RespiBAN and Empatica E4 wearable devices along with machine learning algorithms such as Decision Trees (DT), Random Forests (RF), Linear Discriminant Analysis (LDA), k-Nearest Neighbors (kNN), and AdaBoost. They achieved a level of accuracy of 76.50%. Chakraborty et al. [6] used RespiBAN and Empatica E4 wearables, applying a deep learning Convolutional Neural Network (CNN) model. They achieved an accuracy of 77.06%. On the other hand, Tanwar et al.

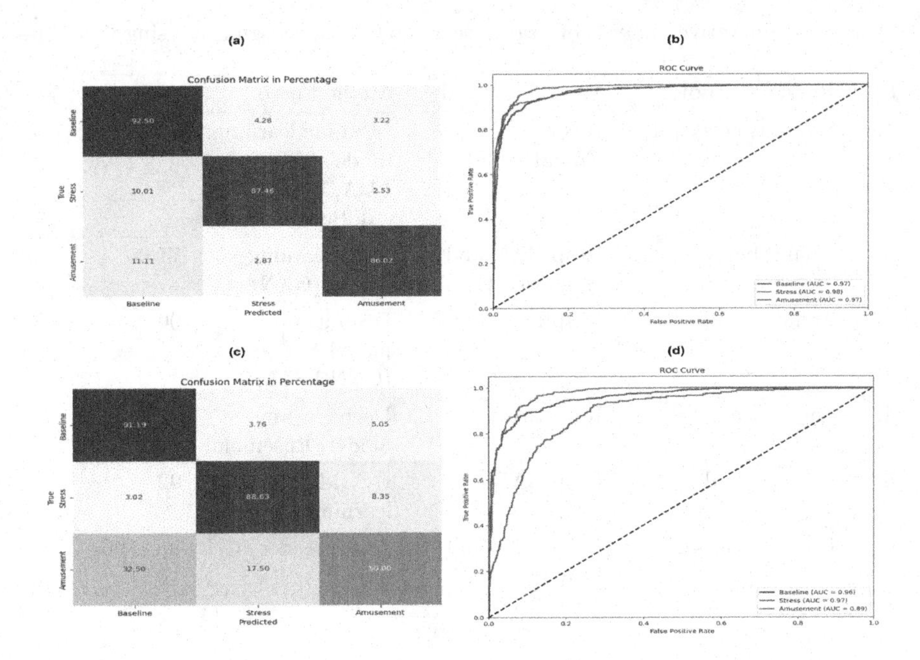

Fig. 5. Confusion matrix (a) and ROC curve (b) obtained with the proposed work using chest wearable and confusion matrix (c) and ROC curve (d) using wrist wearable for 3-class stress classification

[15] utilized a CNN-LSTM based deep learning model using RespiBAN data and achieved a better accuracy of 90%. Singh et al. [13] obtained a 91% accuracy by using a deep learning based ensemble model with RespiBAN data. Tanwar et al. [14] conducted a study where they used a hybrid deep learning model and achieved an accuracy of 92% by using RespiBAN wearables. The proposed approach utilizes RespiBAN and Empatica E4 wearables, implementing an autoencoder based technique. It achieves an exceptional accuracy of up to 96%, showcasing its outstanding performance in stress recognition tasks.

4 Discussion

The integration of AI in the healthcare sector, as outlined in this study, carries substantial ramifications for sustainability, specifically in the context of the provision of healthcare services. The major objective of this study pertains to the progression of AI technologies in the context of diagnostic applications. However, it is imperative to acknowledge the implications of these developments in relation to wider sustainability objectives. The use of AI into healthcare processes has the potential to significantly contribute to the improvement of efficiency and sustainability within healthcare systems. An essential element of our contribution is the optimization of energy efficiency. Through the implementation of automated

Table 3. Comparative analysis of prior research on stress recognition using wearables

S.No.	Previous Studies	Wearables used	Method used	Accuracy (%)
1.	Schimdt et al. [11]	RespiBAN and Empatica E4	Machine learning models DT, RF, LDA, kNN AdaBoost	76.50
2.	Chakraborty et al. [6]	RespiBAN and Empatica E4	Deep learning model (CNN)	77.06
3.	Tanwar et al. [15]	RespiBan	Deep learning model (CNN-LSTM)	90
4.	Singh et al. [13]	RespiBAN	Deep learning model (Ensemble)	91
5.	Tanwar et al. [14]	RespiBAN	Hybrid deep learning model	92
6.	Proposed model	RespiBAN and Empatica E4	Autoencoder (CNN based)	upto 96

and streamlined diagnostic procedures, the suggested AI application effectively mitigates the conventional energy consumption associated with these processes. The utilization of automated diagnostics mitigates the need for extensive manual intervention, resulting in expedited diagnostic processes and reduced operational carbon emissions inside healthcare establishments. The observed decrease in energy consumption is consistent with international endeavors to advance sustainable methodologies in the healthcare sector. Furthermore, this study is in accordance with many Sustainable Development Goals (SDGs). Significantly, it contributes to the achievement of SDG 3 [17]: Good Health and Well-being through the enhancement of healthcare service quality and accessibility. The utilization of AI in diagnostics has the potential to facilitate the timely identification of diseases and improve patient outcomes, hence augmenting public health. Furthermore, through the facilitation of innovation within the healthcare sector, this endeavor makes a valuable contribution to the achievement of SDG 9 [17]: Industry, Innovation, and Infrastructure. It actively promotes the advancement of robust and enduring healthcare infrastructure. Furthermore, the prioritization of resource optimization and responsible consumption is in accordance with SDG 12 [17]: Responsible Consumption and Production, hence reinforcing the significance of this research in shaping sustainable development.

The incorporation of these factors into the analysis not only propels the progress of AI in the healthcare sector but also showcases its congruence with wider sustainability goals, underscoring its potential influence on the trajectory of sustainable healthcare in the future.

4.1 Limitations

Although the proposed AI-assisted diagnostic system demonstrates a notable level of accuracy, it is crucial to acknowledge and address several potential limitations and challenges in order to ensure its practical applicability and reliability in real-world scenarios. Signal interference has the potential to adversely affect the integrity of physiological signals, hence increasing the likelihood of inaccurate measurements. Furthermore, it is important to consider that sensor performance and the overall accuracy of the system might be influenced by shifting external conditions, including fluctuations in temperature and humidity. Moreover, a complete evaluation of the system's long-term performance over longer duration remains incomplete, hence prompting concerns over its resilience and reliability throughout prolonged utilization. In order to ensure the system's robustness and reliability in many real-world settings, it is imperative to consider and address these problems in future study.

5 Conclusion

The proposed study effectively developed a stress recognition model that performed well and had high accuracy. Implementing an autoencoder for feature extraction has demonstrated its efficacy in extracting significant characteristics from the data. The model demonstrated high accuracy, precision, recall, and F1-scores for both chest and wrist wearables in the 2-class stress classification task. Moreover, the model demonstrated a 96% accuracy when utilizing chest wearables and a 93% accuracy when utilizing wrist wearables. The model demonstrated strong performance in 3-class classification, with an accuracy of 84% with chest wearables and 90% with wrist wearables. In addition, the analysis of the confusion matrix and ROC curve highlighted the model's reliability and predictability, as indicated by high values of area under the ROC curve, which demonstrate outstanding discriminatory capacity. The proposed approach demonstrated higher performance in wearables-based stress recognition tasks compared to earlier studies, obtaining an accuracy of up to 96% in comparative analysis.

References

1. Aristizabal, S., et al.: The feasibility of wearable and self-report stress detection measures in a semi-controlled lab environment. IEEE Access **9**, 102,053–102,068 (2021)
2. Bolpagni, M., Pardini, S., Dianti, M., Gabrielli, S.: Personalized stress detection using biosignals from wearables: a scoping review. Sensors **24**(10), 3221 (2024)
3. Bota, P.J., Wang, C., Fred, A.L., Da Silva, H.P.: A review, current challenges, and future possibilities on emotion recognition using machine learning and physiological signals. IEEE Access **7**, 140,990–141,020 (2019)
4. Can, Y.S., Arnrich, B., Ersoy, C.: Stress detection in daily life scenarios using smart phones and wearable sensors: a survey. J. Biomed. Inform. **92**, 103,139 (2019)

5. Can, Y.S., Chalabianloo, N., Ekiz, D., Ersoy, C.: Continuous stress detection using wearable sensors in real life: algorithmic programming contest case study. Sensors **19**(8), 1849 (2019)
6. Chakraborty, S., Aich, S., Joo, M.i., Sain, M., Kim, H.C.: A multichannel convolutional neural network architecture for the detection of the state of mind using physiological signals from wearable devices. J. Healthc. Eng. **2019** (2019)
7. Chen, S., Guo, W.: Auto-encoders in deep learning–a review with new perspectives. Mathematics **11**(8), 1777 (2023)
8. Gedam, S., Paul, S.: A review on mental stress detection using wearable sensors and machine learning techniques. IEEE Access **9**, 84045–84066 (2021)
9. Giannakakis, G., Grigoriadis, D., Giannakaki, K., Simantiraki, O., Roniotis, A., Tsiknakis, M.: Review on psychological stress detection using biosignals. IEEE Trans. Affect. Comput. **13**(1), 440–460 (2019)
10. Machiraju, S.S., Konijeti, N., Batchu, A., Tata, N.: Stress detection using adaptive threshold methodology. In: 2020 5th International Conference on Communication and Electronics Systems (ICCES), pp. 889–894. IEEE (2020)
11. Schmidt, P., Reiss, A., Duerichen, R., Marberger, C., Van Laerhoven, K.: Introducing WESAD, a multimodal dataset for wearable stress and affect detection. In: Proceedings of the 20th ACM International Conference on Multimodal Interaction, pp. 400–408 (2018)
12. Schmidt, P., Reiss, A., Duerichen, R., Van Laerhoven, K.: Wearable affect and stress recognition: a review. arXiv preprint arXiv:1811.08854 (2018)
13. Singh, G., Phukan, O.C., Kumar, R.: Stress recognition with multi-modal sensing using bootstrapped ensemble deep learning model. Expert Syst. **40**(6), e13,239 (2023)
14. Tanwar, R., Phukan, O.C., Singh, G., Pal, P.K., Tiwari, S.: Attention based hybrid deep learning model for wearable based stress recognition. Eng. Appl. Artif. Intell. **127**, 107,391 (2024)
15. Tanwar, R., Phukan, O.C., Singh, G., Tiwari, S.: CNN-LSTM based stress recognition using wearables. In: KGSWC Workshops, pp. 120–129 (2022)
16. Tanwar, R., Singh, G., Pal, P.K.: Fuser: Fusion of wearables data for stress recognition using explainable artificial intelligence models. In: 2023 14th International Conference on Computing Communication and Networking Technologies (ICCCNT), pp. 1–6. IEEE (2023)
17. Vinuesa, R., et al.: The role of artificial intelligence in achieving the sustainable development goals. Nat. Commun. **11**(1), 1–10 (2020)

Optimizing Agricultural Practices Through Integrated IoT and ML Solutions

Yadidiah Kanaparthi, Abdul Karim Shaikh, Inaya Imtiyaz Khan, and Rita Zgheib(✉)

Computer Science Department, Canadian University Dubai, Dubai, UAE
Rita.zgheib@cud.ac.ae

Abstract. The integration of the Internet of Things (IoT) and Machine learning (ML) techniques can enhance agricultural practices by providing real-time data and insights into the implementation of these techniques and the risks associated with them. The paper answers questions on data acquisition and farming optimization by equipping farms with sensors and analysing the data through data sets. It also focuses on crop recommendation, yield prediction, and weather forecasting, demonstrating how the integration of IoT and ML provides optimized solutions for decision-making, environmental sustainability, and improved crop yield. The analysis presents a variety of ML algorithms such as the Naïve Bayes classifier, Decision Tree, Random Forest, K-Nearest Neighbours (KNN), Regression techniques, and Support Vector Machine (SVM) that have been used to assess the ensemble techniques to evaluate their effectiveness using model accuracy and F1 score in optimizing agricultural processes. Six different data sets have been used to cover all possible scenarios. The findings highlight the potential of IoT-ML integration to improve decision-making, environmental sustainability, and crop yield. Overall, this research contributes to the advancement of smart agriculture and offers actionable insights for stakeholders in the agricultural sector.

Keywords: IoT · Machine Learning · sustainability · agriculture practices

1 Introduction

The Sustainable Development Goals (SDGs), established by the United Nations, aim to address global challenges and guaranteeing a better and more sustainable future for everybody. SDG 2: Zero Hunger aims to eliminate hunger, secure food security, enhance nutrition, and advance sustainable agriculture by 2030. Achieving this ambitious goal requires innovative solutions that leverage advanced technologies to optimize food production, distribution, and consumption processes.

The Internet of Things (IoT) and Machine Learning (ML) are two such technologies with significant potential to contribute to SDG 2. IoT enables the seamless integration of various devices and systems, making it easier to monitor and manage agricultural operations in real-time. Examples of IoT applications that can increase crop productivity, decrease waste, and guarantee resource efficiency include sensors, drones, and smart irrigation systems. On the other hand, ML algorithms have the ability to examine enormous

S. Tiwari et al. (Eds.): AI4S 2024, CCIS 2243, pp. 89–103, 2025.
https://doi.org/10.1007/978-3-031-81369-6_8

volumes of data produced by Internet of Things (IoT) devices in order to offer useful insights, forecast crop diseases, enhance supply chains, and assist in decision-making. To further elucidate the organization of a typical IoT framework and its integration with innovative techniques across various domains, we highlight the importance of a structured approach as demonstrated in [22], where a hybrid intelligent system is effectively utilized for channel allocation and packet transmission in CR-IoT networks, showcasing the potential of IoT in enhancing system efficiency and adaptability.

The world population is increasing drastically which will require more agricultural products. According to data from the Food and Agriculture Organization (FAO) of the United Nations, the world's population will increase to 9.8 billion people by 2050, and an additional 50% of the current food production is needed for this population increase. Artificial Intelligence (AI) can be used to improve crop planting efficiency, by improving the quality and quantity of the crops [1]. IoT systems like drones can make it easier to produce crops more efficiently as drones can plant more than 500 seeds of crop per hour compared to 800 seeds planted by farmers per day [2]. These technical advancements not only increase productivity on the farm but also reduce farmers' workload as well as reduce labour costs while also providing detailed monitoring analytics of their produce [3]. Using IoT has made a positive impact in agriculture research in this field is continually expanding, Farmers can maximize their land's potential by using this technology. IoT uses site-specific strategies to protect the environment and promote sustainability. According to expert forecasts, the global smart farming market is expected to experience significant growth, with projections indicating a rise from $14.65 billion in 2021 to $66 billion by 2030, corresponding to a compound annual growth rate (CAGR) exceeding 18% during the period from 2022 to 2030 [12]. IoT makes it easier to collect and process data collected from sensors to make detailed and visual analyses, IoT makes remote farm monitoring possible allowing farmers to have their crop information from any location at any time. Predictive analytics algorithms can be used with data from sensors to estimate variables like crop productivity and the best times to plant and harvest [13]. With the help of these predictions, farmers can make informed choices that can boost crop quality, productivity, and production.

Precision farming maximizes resource efficiency while minimizing waste and environmental effects by precisely controlling resources such as water, fertilizers, and pesticides based on real-time data and forecast insights. By lowering input prices, this strategy helps farmers both financially and in terms of sustainable farming practices. The integration of IoT in agriculture transforms how farmers manage their operations and provides several benefits. IoT-ML-based agriculture is the next evolutional thing in smart agriculture. Applying ML to data generated from various IoT devices can help provide definitive information and richer insights for crop production improvements [4]. Ultimately, IoT in agriculture is a key driver for achieving sustainability, increasing productivity, and meeting the challenges posed by a growing global population, farmers can increase crop yields, streamline processes, and guarantee the long-term sustainability of food production systems in a rapidly evolving environment [14]. The crop recommendation model suggests the best crops to cultivate based on factors like soil type, weather conditions, and historical data, aiming to optimize yield and resource use. The yield prediction model estimates the expected crop output by analyzing variables such as climate, soil quality, and farming practices. The weather forecasting model predicts short- and long-term weather conditions, helping farmers make informed decisions. Based on this introduction This paper aims to find how the IoT obtained data can be used to get detailed farm analysis and crop improvement with the help of ML and We discuss how these technologies can be harnessed to address the challenges of hunger and food insecurity in the context of SDG 2. This study also evaluates the effectiveness of ensemble learning techniques and compares them with plain ML model results.

2 Literature Review

Research aligning IoT and ML with SDG 2 highlights their critical role in addressing hunger and food insecurity. For instance, IoT-enabled precision agricultural methods may improve crop yields and optimize resource use, directly supporting SDG 2 objectives. To guarantee food security in vulnerable areas, machine learning algorithms can also be used to improve distribution networks and anticipate food shortages. For instance, ML algorithms can evaluate vast amounts of data collected by IoT sensors on crop health, weather patterns, and soil conditions to produce useful insights for improving agricultural operations [20].

Furthermore, research indicates that the application of IoT and ML in agriculture could drastically reduce post-harvest losses, a key problem that causes food scarcity. These technologies can support preserving the quality and safety of food that has been stored by keeping an eye on storage conditions and identifying possible problems, which is in line with SDG 2 goals [21].

Sarkar et al. [1] explored the impact of AI in crop automation and found Robotics have revolutionized various aspects of crop management harvesting, growth monitoring, picking, sorting, and packing. Drones play a major role in managing many operations like spraying, monitoring crop health, disaster management, and soil analysis with the help of Advanced computer vision [5], But adoption of these innovative technologies is still less due to a lack of awareness, and skills to utilize these technologies, additionally implementing robotics in farming increases costs and complexity issues leading to low IoT adoption. A paper by Han [6] has a similar result which argues that the lack of awareness prevents the implementation of these systems and mentions that implementing technology in agriculture has benefits for both producers and consumers.

According to Aggarwal and Singh [7], farmers deal with a variety of challenges in agriculture, including irrigation management, analyzing soil behaviour, predicting crop development, and managing disease. Discusses weather forecasting plays an important role in addressing these challenges. Temperature fluctuations, precipitation patterns, and other meteorological phenomena directly affect the output of agriculture and play an important part in tackling these issues.

Through the utilization of AI and IoT technology, farmers can obtain predictive analytics and real-time weather data to maximize crop yields, reduce resource consumption, and lessen the negative consequences of extreme weather occurrences. Adoption of weather forecasting technologies is essential for ensuring sustainable agricultural practices and enhancing food security in the face of changing climatic conditions. [8] Also, similar results found that in almost all fields of agriculture such as farm monitoring, irrigation, and pest monitoring IoT techniques can be applied and the IoT system needs to integrate with other technologies like machine learning to deal with the vast amounts of agricultural data.

Benos et al. [9] conducted a review and found that most of the articles have primarily focused on crop management with an increase in crop recognition, ML algorithms particularly Artificial Neural Networks (ANNs), have been widely utilized for handling heterogeneous data in agriculture there is also a growing interest for Ensemble Learning (EL) and Support Vector Machines (SVMs) demonstrate high accuracy for data analysis.

Gia et al. [10] argue that relying solely on traditional IoT architectures cannot guarantee that the systems work properly because cloud-centric IoT applications cannot be implemented in remote areas where the Internet is not stable, or coverage is limited. In these situations, data real-time data monitoring and processing is not possible, However, effective solutions like Edge and Fog computing offer numerous benefits including effective sensor data, reduced network load and the ability to be executed by small microcontrollers such as Raspberry Pi, Arduino, or ESP32.

The paper [11] emphasizes the importance of forecasting crop performance under various environmental conditions to enhance crop productivity and argues that existing IoT platforms are not designed to support near-real-time analysis of large data gathered from sensors. Thus, highlighting the need for innovative smart farming solutions.

According to [19] Ensemble learning enhances model accuracy and resilience by merging predictions from multiple models. It also mitigates biases that exist in individual models by leveraging the collective intelligence of the ensemble. Stacking, where predictions from multiple models are used to build a new model. Blending, where a holdout set is derived from the training set to make predictions. Bagging, where the results of multiple models are combined to get a generalized result, and boosting where each subsequent model attempts to correct the errors of the previous model are the main ensemble techniques. And their effectiveness is evaluated with the general model.

3 Methodology

In this project, datasets collected from several different platforms such as Kaggle and ACM have been used for analysis. After data preprocessing which included handling outliers, missing values, dimensionality reduction, and data integration, a total of six datasets were finalized to be used for different functionalities of the research process. Each dataset was divided into a split of 70/30 as a training and testing set for model development and evaluation. The below Table 1 list the details of the six datasets used for this study.

The first dataset contains the following attributes.

- N - ratio of Nitrogen content in the soil
- P - ratio of Phosphorous content in soil
- K - ratio of Potassium content in soil
- temperature - temperature in degrees Celsius
- humidity - relative humidity in %
- ph - ph value of the soil

Table 1. Details of the datasets used in the study.

ID	Name	Target	Samples	Features
1	Crop_recommentation	Factors affecting crop	2201	8
2	DailyDelhiClimateTrain	Factors affecting climate	1463	5
3	pesticides	Usage of pesticides country-wise	4350	7
4	rainfall	Average rainfall country-wise	6728	3
5	temp	Average temperature year over the past few years	71312	12
6	yield	Yield produced in each country yearly	56718	

- rainfall - rainfall in mm

The following attributes N, P, and K can be collected by soil NPK sensor [16], the temperature and humidity can be collected using the DHT 11 sensor, the Ph sensor can measure the acidity of liquids, rainfall can be measured by a Rain Guage in [15].

Table 2. Excerpt of the dataset.

P	K	Temperature	Humidity	Ph	Rainfall	Label
35	25	24.0	61.1	7.0	161.5	Coffee
57	53	42.3	90.5	6.9	74.9	Papaya
57	41	21.4	84.9	5.8	272.2	Rice
126	204	23.1	23.1	92.8	6.4	Apple
67	22	69.4	69.4	6.6	51.6	Lentil

Table 2 shows how the sample is structured, here the values are rounded to 1 decimal place. A ML model was developed for crop recommendation which uses the best ML algorithm based on the accuracy score and its F1 score. Then 4 ensemble models Stacking, Blending, Bagging, and Boosting are trained on the base models to find the best model to accurately predict the optimal type of crop to plant based on the given environment conditions. For the base model evaluation, various machine learning algorithms like decision tree, SVM, random forest, K-Nearest Neighbours (KNN), Naïve Bayes, linear regression, and logistic regression were used, after evaluating the optimal model based on various scores, the model with the highest accuracy, precision, and F1 score was selected.

The next task was crop yield prediction, Here the datasets rainfall, yield, pesticides, and temp were combined in the yield dataset to create a data frame for predicting yield the data frame was then encoded from categorical to numerical data, test and training sets were divided and a comparison between actual yield and predicted yield was formed.

Further analysis was performed over the dataset that gave additional valuable information over the merged dataset. Then, the decision tree regressor was used to predict crop yield and ensemble learning techniques that were trained on various models including ridge, Decision Tree, Random Forest Regressor, AdaBoost Regressor, and Gradient Boosting Regressor. The best model was compared using R-squared values.

Next, for predicted weather forecasting over the time series dataset here are the following attributes.

- date - format YYYY-MM-DD
- meantemp - Mean temperature from daily 3-h intervals.
- humidity - Humidity value for the day per cubic meter volume of air.
- wind_speed - measured in km/h.
- meanpressure - Pressure reading of weather measure in atm.

The Adafruit Anemometer Sensor can be used for measuring wind speed [17], and the DCT 532 sensor can be used for measuring pressure readings. Data Preprocessing was done over the dataset the ARIMA model was used for time series data forecasting [18] and the model score with the help of evaluation metrics like Mean Squared Error. The ensemble learning models that were used were Stacking Regressor, RandomForestRegressor, RidgeCV, Decision Tree, Bagging Regressor, and XGBRegressor (XGBoost) multiple evaluation metrics were used like Mean Absolute Percentage Error (MAPE), Mean Squared Error (MSE), Root Mean Squared Error (RMSE), and Akaike Information Criterion (AIC). The algorithm used in this study is described below.

```
Crop_Recommendation ():
    data = collect_data()
    data_preprocessing = preprocess_data(data)
    training_data, testing_data = split_data(preprocessed_data)
    dataset = create_dataset(training_data)
    model = develop_model(dataset)
    evaluation_results = evaluate_model(model, testing_data)
    print_results(evaluation_results)
```

4 Results

The data from the dataset provides valuable information for optimizing farming practices, and evaluating the prediction model provides more valuable information so that intelligent decisions can be made using the results.
Crop Recommendation Dataset:
This dataset stores the values that can be used to predict crop types based on the inputs. The scatter plot describes plots of all the different crops based on the two most important features, temperature and rainfall.

The scatter plot in Fig. 1 indicates, most of the crops have the optimal temperature of 20–30 °C and rainfall from 50–200 mm. Rice seems to be the only crop that needs rainfall over 200 mm. Papaya can be grown in high temperatures ranging from 2545 °C,

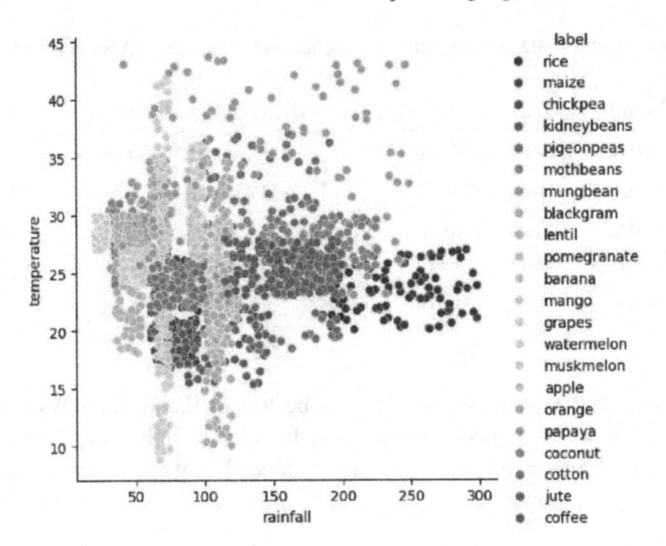

Fig. 1. Scatter Plot

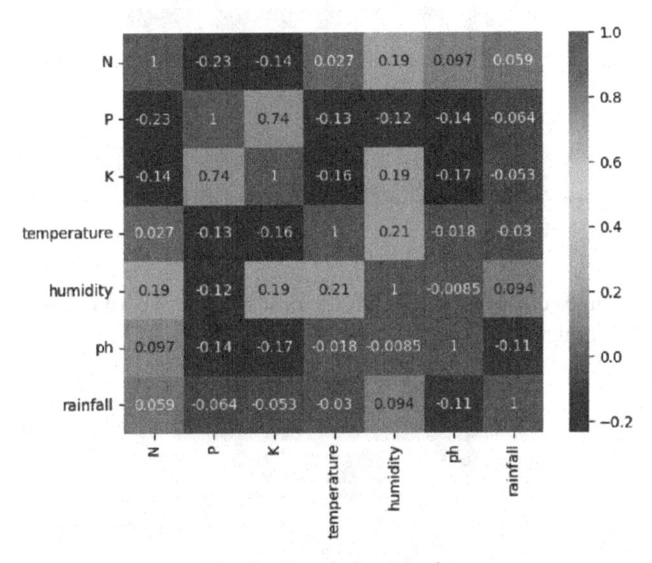

Fig. 2. Correlation Matrix

and grapes and oranges can be grown in low temperatures. Although, grapes seem to grow in all temperatures ranging from 10–45 °C.

The correlation matrix in Fig. 2 indicates that attributes with the highest correlation are phosphorus and potassium, pH and Nitrogen, and humidity and rainfall. Rainfall and nitrogen seem to have a high correlation. This correlation matrix is used to choose the best attributes for feature selection. Comparing various crop recommendations gives the following results.

Table 3. Comparative Data on Agricultural Practices and Outcomes Across Different Regions

Area	Item	Year	Rainfall (mm)	Pesticide (Tonn)	Temp (°C)	Yield (hg/ha)
Zambia	Maize	2004	1020.0	1670.0	20.79	19239
Colombia	Yams	2004	3240.0	1.1 × 105	27.3	117050
Guyana	Maize	1996	2387.0	289.9	27.11	11481
India	Potatoes	1991	1083.0	72133.0	25.99	162540
Guinea	Plantains	2010	1651.0	556.24	27.8	52256

Evaluating Table 3, valuable insights can be found like all models except the linear regression model had good accuracy and F1 scores. Indicating linear regression should not be used for classification tasks as evidenced by its poor performance metrics compared to the other models.

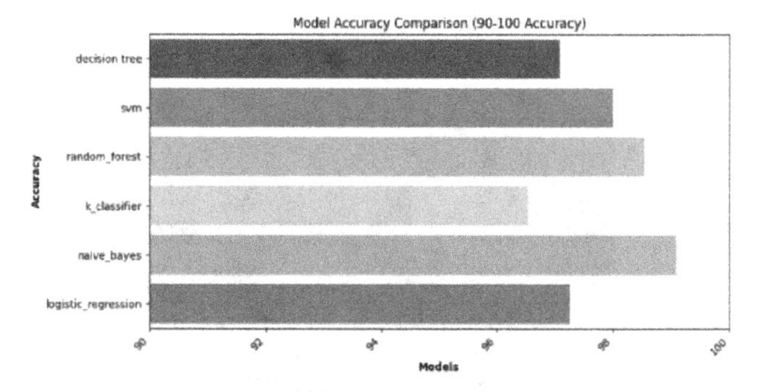

Here Naïve Bayes classifier model has the best accuracy score followed by Random Forest and SVM models.

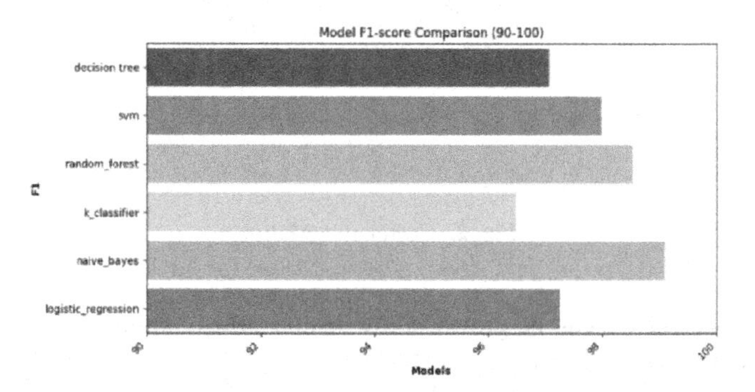

Here again Naïve Bayes classifier model has the best F1 score followed by Random forest and SVM models. The results found that the Naïve Bayes classifier model works well with data for Crop recommendation and works faster than other models as this model can be run in 1.6 ms. Naïve Bayes is known for its robustness against overfitting, which likely played a significant role in maintaining consistent performance across various subsets of our data. This resilience ensures that the model remains reliable and effective, even when faced with new or varied data inputs, further emphasizing its value in our application.

Comparing Naïve Bayes accuracy with the best accuracy from the output of the ensemble model gives the below graphs.

Model	Accuracy	F1 score	Test time
Decision Tree	0.9709	0.9708	0.0013
SVM	0.9800	0.9798	0.0179
Random Forest	0.9855	0.9853	0.0037
KNN	0.9655	0.9649	0.0056
Naïve Bayes	0.9909	0.9908	0.0016
Linear Regression	0.0636	0.0464	0.0015
Logistic Regression	0.9727	0.9727	0.0019

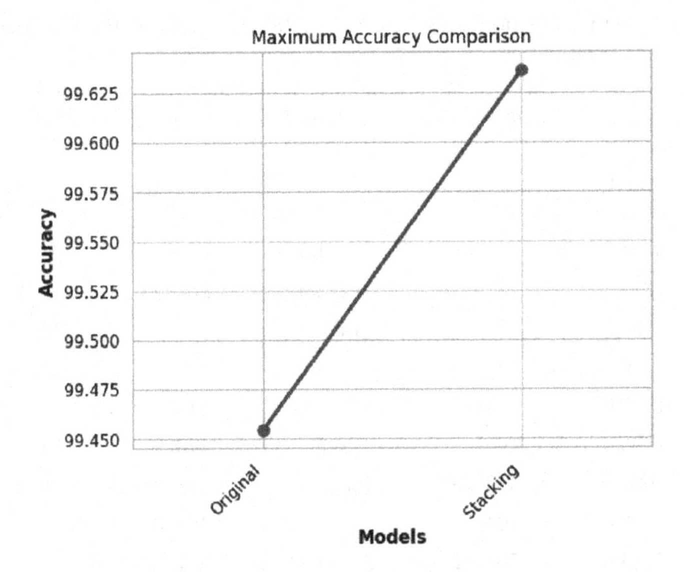

The point plot above shows that the stacking ensemble model has better accuracy than the naïve bayes score. The f1 score also gives the same idea.

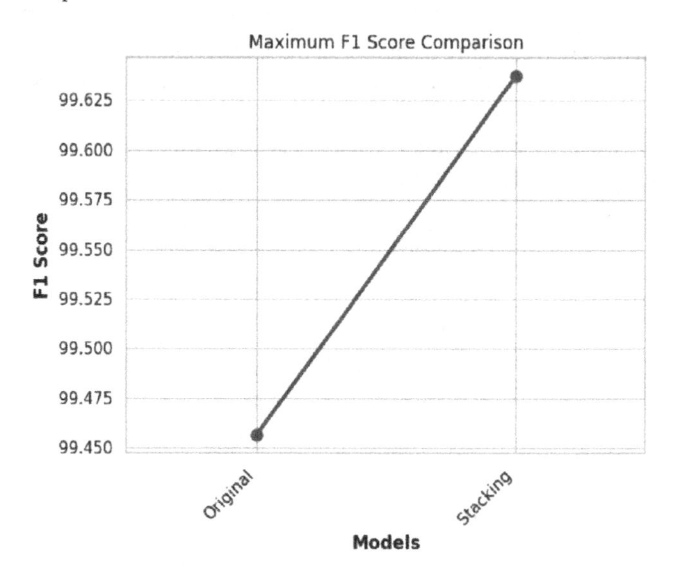

Comparing the results indicates that the stacking model has the best performance for crop recommendation.

So far Naïve Bayes classifier model had the highest accuracy score and the F1 score, however, the results indicate that the stacking model also has the best performance for crop recommendation.

Crop Yield prediction:

Index	Year	Yield	Avg. Rain	Pest	Avg. Temp
count	28242.0	28242.0	28242.0	28242.0	28242.0
mean	2001.5	77053.3	1149.06	37077	20.54
std	7.05	84956.6	709.81	59958.8	6.31
min	1990.0	50.0	51.0	0.04	1.3
25%	1995.0	19919.3	593.0	1702.0	16.7
50%	2001.0	38295.0	1083.0	17529.4	21.51
75%	2008.0	104677	1668.0	48688	26.0
max	2013.0	501412	3240.0	367778	30.65

Note: Avg = Average, Rain = Average Rainfall, Pest = Pesticides Tones, Temp = Average Temperature The dataset directly provides valuable insights like the total amount of data collected (over 28,000 rows), the lowest yield (50 hg/ha), and the highest (501412.0 hg/ha). On average, around 37,000 tonnes of pesticides were used, and the average temperature of the location was around 20.54 °C.

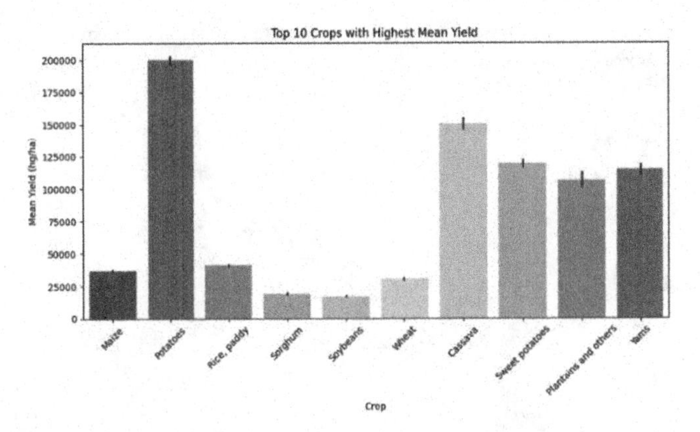

The graph shows that potatoes are the highest-grown crop followed by Cassava, Sweet potatoes, and Yams. After the analysis process, the decision tree regressor model provided the following results.

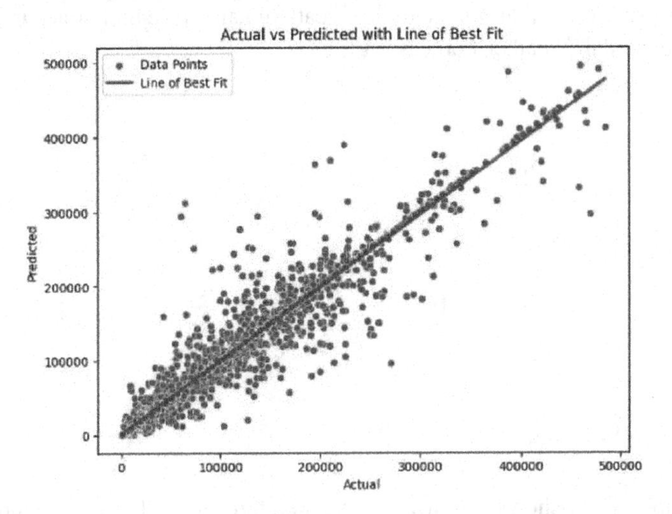

The graph shows that as the actual values increase, the predicted values also tend to increase, also the values seem to be closer to the line of best fit, further evaluation of the model shows that R-squared: 0.961, this shows a strong positive relationship between both values, so the model was successful in determining yield prediction. Also, the ensemble model gives much better results for R-squared (0.978) the line of best fit for bagging gives a comparison of the accuracy of the decision tree model.

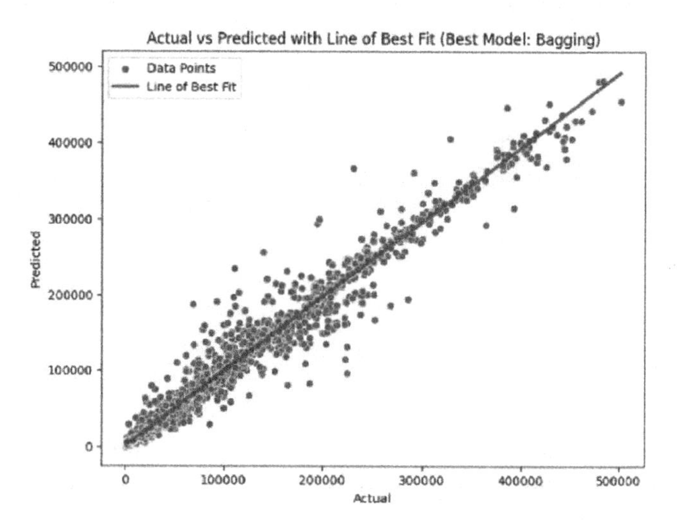

The graph indicates that points are less scattered and have higher accuracy than the decision tree model. Weather Forecasting:

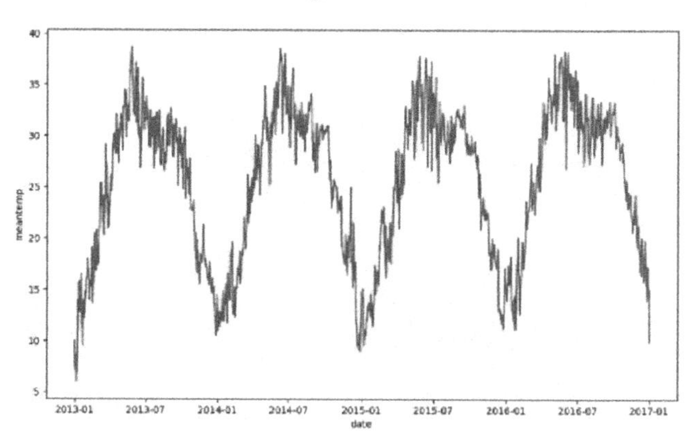

The graph above shows, the average temperature in the location where data was collected was from 2013 Jan to 2017 Jan. Using the ARIMA model gives forecasting predictions.

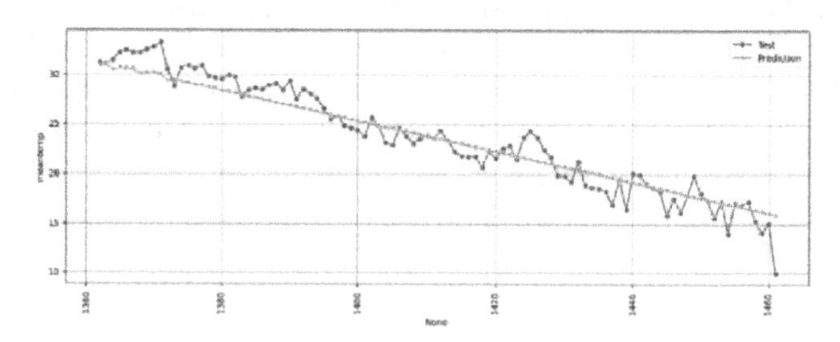

The table below gives all the information regarding the evaluation metrics.

Metric	ARIMA	Stacking	Blending	Bagging	Boosting
Mape	0.056	0.093	0.118	0.093	0.081
MSE	2.277	8.513	9.244	9.113	6.453
RMSE	1.509	2.918	3.040	3.019	2.540
AIC	5469.426	5469.426	5469.426	5469.426	5469.426

Evaluation metrics.
Note: MAPE = Mean Absolute Percentage Error, MSE = Mean Squared Error, RMSE = Root Mean Squared Error, AIC = Akaike Information Criterion.

Although some variation is shown between the test and prediction values, this model was still accurate as the Mean Absolute Percentage Error was 0.0559, Mean Squared Error was 2.277 and Root Mean Squared Error was 1.508 with ARIMA having the best results. The final model which is ARIMA has a fairly low error compared to the actual temperature.

5 Conclusion

In Conclusion, addressing the challenge of feeding a growing population, innovative approaches are necessary for agricultural productivity and sustainability, This research utilized IoT and ML technologies in addressing these challenges, and offered tangible solutions to optimize crop yields, enhance resource efficiency, and promote sustainable monitoring.

Key findings from this research show the effectiveness of ensemble learning techniques in optimizing agricultural-related tasks. Models such as stacking and bagging showcase enhanced accuracy and resilience, paving the way for more informed and efficient farming practices.

Furthermore, the ARIMA model's ability to provide useful insights for time series data is shown by its accuracy in weather forecasting. Farmers can manage irrigation, pest control, and planting schedules to maximize yield and minimize resource waste by precisely forecasting weather patterns.

Although the paper covers the majority of the aspects, there are still some limitations of the present study such as the evaluation metrics used do not encompass the entire spectrum of performance indicators relevant to agricultural decision-making and also, as the study was primarily focused on prediction and optimization, economic and social implications may need to be discussed in detail.

Ultimately, this paper contributes valuable insights into the transformative potential of IoT and ML in agriculture, facilitating the achievement of SDG 2. It opens the door to more informed and efficient farming methods amidst evolving economic and environmental circumstances, it provides access to more knowledgeable and effective farming techniques, assuring a sustainable and food-secure future for everybody.

The field of agriculture has a multitude of fascinating prospects in research and development. Implementing deep learning methods like convolutional neural networks would further refine the results and utilizing ensemble learning methodologies with added features could improve the results further. For instance, these networks could be utilized to process intricate image data, like satellite images, to enhance crop management strategies. The integration of various models can improve both predictive accuracy and overall robustness. Additionally, a more efficient data processing approach can reduce processing time by merging edge and fog computing technologies. Finally, new, and innovative methods need to be found to develop more efficient agricultural practices.

References

1. Sarkar, Md.R., Masud, S.R., Hossen, M.I., Goh, M.: A comprehensive study on the emerging effect of artificial intelligence in agriculture automation. In: 2022 IEEE 18th International Colloquium on Signal Processing & Applications (CSPA), pp. 419–424. IEEE, Selangor, Malaysia (2022). https://doi.org/10.1109/CSPA55076.2022.9781883
2. Tjhin, V.U., Riantini, R.E.: Smart farming: implementation of industry 4.0 in the agricultural sector. In: 2022 6th International Conference on E-Commerce, E-Business and E-Government, pp. 115–120. ACM, Plymouth United Kingdom (2022). https://doi.org/10.1145/3537693.3537711
3. Sheham, Md.N.H., et al.: Design of an IoT-based Smart Farming System: IoT-based Smart Farming System. In: Proceedings of the 2nd International Conference on Computing Advancements, pp. 284–293. ACM, Dhaka Bangladesh (2022). https://doi.org/10.1145/3542954.3542996
4. Maduranga, M.W.P., Abeysekera, R.: Machine learning applications in iot based agriculture and smart farming: a review. Int. J. Eng. Appl. Sci. Technol. **04**(12), 24–27 (2020). https://doi.org/10.33564/IJEAST.2020.v04i12.004
5. Said Mohamed, E., et al.: Smart farming for improving agricultural management. Egypt. J. Remote Sens. Space Sci. **24**(3), 971–981 (2021). https://doi.org/10.1016/j.ejrs.2021.08.007
6. Han, D.: Big Data Analytics, Data Science, ML&AI for Connected, Data-driven Precision Agriculture and Smart Farming Systems: Challenges and Future Directions. In: Proceedings of Cyber-Physical Systems and Internet of Things Week 2023, pp. 378–384. ACM, San Antonio TX USA (2023). https://doi.org/10.1145/3576914.3588337
7. Aggarwal, N., Singh, D.: Technology assisted farming: Implications of IoT and AI. IOP Conf. Ser. Mater. Sci. Eng. **1022**(1), 012080 (2021). https://doi.org/10.1088/1757899X/1022/1/012080
8. Terence, S., Purushothaman, G.: Systematic review of Internet of Things in smart farming. Trans. Emerg. Telecommun. Technol. **31**(6), e3958 (2020). https://doi.org/10.1002/ett.3958
9. Benos, L., Tagarakis, A.C., Dolias, G., Berruto, R., Kateris, D., Bochtis, D.: Machine learning in agriculture: a comprehensive updated review. Sensors **21**(11), 3758 (2021). https://doi.org/10.3390/s21113758
10. Gia, T.N., et al.: Edge AI in Smart Farming IoT: CNNs at the Edge and Fog Computing with LoRa
11. Jayaraman, P., Yavari, A., Georgakopoulos, D., Morshed, A., Zaslavsky, A.: Internet of things platform for smart farming: experiences and lessons learnt. Sensors **16**(11), 1884 (2016). https://doi.org/10.3390/s16111884
12. Sydoruk, A.: 10 Key Benefits of IoT in Agriculture and Farming. smarttek.solutions. https://smarttek.solutions/blog/iot-in-agriculture/. Accessed 26 Feb. 2024
13. Lerner, M.: How IoT Precision Agriculture And Smart Farming Works. ridge.co. https://www.ridge.co/blog/how-iot-precision-agriculture-and-smart-farming-works/. Accessed 26 Feb. 2024
14. Puzhevich, V.: The benefits of IoT devices in agriculture. scand.com. https://scand.com/company/blog/the-benefits-of-iot-in-agriculture-infographic/. Accessed 26 Feb. 2024

15. Aggiustatutto, G.: DIY Arduino Rain Gauge. instructables.com. https://www.instructables.com/DIY-Arduino-Rain-Gauge/. Accessed 26 Feb. 2024
16. Alam, M.: Measure Soil Nutrient using Arduino & Soil NPK Sensor. how2electronics.com. https://how2electronics.com/measure-soil-nutrient-using-arduinosoil-npk-sensor/. Accessed 27 Feb. 2024
17. Alam, M.: How to Measure Wind Speed using Anemometer & Arduino. how2electronics.com. https://how2electronics.com/measure-wind-speed-usinganemometer-arduino/. Accessed 27 Feb. 2024
18. Pathak, P.: Building an ARIMA Model for Time Series Forecasting in Python. analyticsvidhya.com. https://www.analyticsvidhya.com/blog/2020/10/how-to-create-anarima-model-for-time-series-forecasting-in-python/. Accessed 27 Feb. 2024
19. Singh, A.: A Comprehensive Guide to Ensemble Learning" analyticsvidhya.com https://www.analyticsvidhya.com/blog/2018/06/comprehensive-guide-for-ensemblemodels/. Accessed 22 April 2024
20. Bongiovanni, R., Lowenberg-Deboer, J.: Precision agriculture and sustainability. Precision Agric. **5**(4), 359–387 (2004)
21. Fanzo, J.: The Nutrition Challenge in Sub-Saharan Africa. UNDP Working Paper (2012)
22. Asuquo, D.E., Umoh, U.A., Robinson, S.A., Dan, E.A., Udoh, S.S., Attai, K.F.: Hybrid intelligent system for channel allocation and packet transmission in CR-IoT networks. Int. J. Hybrid Intel. Sys. **20**(2), 101–117 (2024). https://doi.org/10.3233/HIS-240009

AI in Building Systems for Perpetual Monitoring and Control of ESG Practices

Amit Aylani(✉), Madhuri Rao, and G. T. Thampi

Thadomal Shahani Engineering College, Bandra, Mumbai, India
prof.amitaylani@gmail.com, madhuri.rao@thadomal.org

Abstract. The need for inter-governmental protocols on climate change, world trade agreements, and increased awareness of human rights and sustainability issues has made it imperative for corporations globally to adopt environmental, social, and governance (ESG) practices at a faster rate. This paper highlights the key areas where digital and artificial intelligence can be successfully combined to address these demands. As the importance of ESG and sustainability issues grows, organization are actively seeking out advanced technological solutions to tackle intricate problems. The research methodology used in this paper incorporates a framework, highlighting how technology can address complex issues in data collection and contribute to sustainable practices. The results suggest that AI technology has the potential to greatly improve ESG practices in different areas. It can help organizations enhance their corporate governance by automating compliance monitoring, assessing risks, and providing decision support systems. This paper investigates artificial intelligence technology in creating and implementing an architectural solution for environmental, social, and governance practices, providing a strong foundation for further exploration and application. By using artificial intelligence and IoT, organizations can better align themselves with sustainable development goals and ESG reporting.

Keywords: ESG · Sustainability · AI · Technology solution

1 Introduction

Worldwide, businesses are progressively integrating environmental, social, and governance (ESG) considerations into their operational strategies. ESG is a measure used to assess a company's non-financial performance, focusing on social inequality, corporate governance challenges, climate change, working conditions, and sustainability [6]. These endeavors are in line with global initiatives like paris agreement, which seeks to restrict global temperature rise to no more than 1.5 degrees Celsius. The targets encompass a 45% reduce in emissions by 2030 and the ultimate goal of attaining net-zero emissions by 2050[1].

ESG principles is impacting trade agreements and investment choices on a global scale. This evolve world order is being reshaped by ESG considerations, which are

[1] https://www.un.org/climatechange.

S. Tiwari et al. (Eds.): AI4S 2024, CCIS 2243, pp. 104–114, 2025.
https://doi.org/10.1007/978-3-031-81369-6_9

influencing decision-making in finance, economic policies, and international trade. ESG principles play a crucial role in shaping and transforming of the modern global environment.

An ESG tool which is helpful to assess sustainability practices and performance of an organization. This tool helps in evaluation of potential risks and opportunities related to environmental, social and governance variables while giving the company's general approach to sustainability. Presently, companies have a duty to disclose their ESG metrics in their financial reports and many choose to present them either in their annual or separate sustainability reports. Investors are interested in firms that prioritize ESG principles.

1.1 ESG Factor

ESG factors encompass Environmental, Social, and Corporate Governance conditions that can influence financial performance positively or negatively (Table 1).

Table 1. ESG Factor

Environment Factor	Social Factor	Governance Factor
Greenhouse Gas Emissions Tables	Inequality	Executive remuneration
Waste and Pollution	Health and Safety	Audits and internal controls
Pollution and Waste	Child Labor	Shareholder rights
Climate Change	Human Rights	Anti-competitive practices
Energy Efficiency	Forced Labor	Board Diversity and structure
Greenhouse gas Emission	Employee Benefits	Management structures

1.2 ESG Score

An ESG score evaluates an organization's performance in terms of sustainability metrics related to ESG issues. These scores are generated by rating platforms where analysts evaluate corporate disclosures, and publicly available annual. Various stakeholders, uses these scores for different purposes. An ESG score, along with its components, demonstrates that a company is actively incorporating ESG practices into its operations.

In January 2022, the securities and exchange board of India (Sebi) introduced the business responsibility and sustainability report (BRSR) framework for 1000 listed companies in India. This framework highlights the various metrics of ESG practices, consistent with international standards for sustainability. Besides BRSR, there are some other well-known ESG reporting frameworks available globally, including the Global Reporting Initiative (GRI), the Sustainability Accounting Standards Board (SAAB), the Task Force on Climate-related Financial Disclosures (TCFD), the Carbon Disclosure Project (CDP), Integrated Reporting (IR), the UN Global Compact (UNGC), and the Principles for Responsible Investment (PRI). Such frameworks offer comprehensive guidelines for companies to determine their ESG scores, promoting transparency and accountability in their sustainable practices.

1.3 Digital Technology for Building Efficiencies ESG Practices

Technologies have become part and parcel of ESG practices. The shift from CSR to ESG frameworks has particularly recognized the growing importance of sustainability. However, in the current scenarios, most ESG criteria do not take into account the special impacts of AI, giving companies little motivation to fully analyze and disclose these consequences. This may result in a gap in which significant sustainability-related impacts of AI are overlooked, leaving possible adverse impacts unreported [11].

Integration of technology into ESG processes is essential if accountability and transparency are to be enhanced and as a means of effectively storing, analyzing and reporting data. AI-powered technology helps in simplifying ESG compliance, reducing error rates, and promoting timely reporting. To address sustainability concerns, there must be technological advancements. Digital platforms enable stakeholder engagement through enhanced communication which promotes collaboration as well as trust.

AI framework and technology solutions will be the focus of this paper when it comes to assessing ESG ratings. This includes collecting, analyzing and evaluating ESG variables using cutting-edge technologies like artificial intelligence (AI), internet of things (IoT), natural language processing (NLP) and data analytics. Through NLP and IoT sensors that gather real time information on emissions such as carbon emissions among others, NLP extracts relevant information from annual reports or other documents. In combining these techniques, a large comprehensive dataset has been provided for organizations towards transparent reporting purposes.

2 Literature Review

Wei Wu et al. [3] stated that the challenge is that companies may overstate or slur their actual sustainable performance and the entire ESG reporting process is hidden backward, thereby making the report untrustworthy. They proposed an architecture for smart ESG reporting platform that uses IoT and blockchain technologies. This platform enables corporations the ability to measure and track occupancy levels for environmental data and enhance the security, transparency and creditability of ESG reporting process by focusing on large amounts of raw data.

Akshat Gupta et al. [4] proposed method in two steps for integrating ESG and financial parameters of publicly listed companies. First created a dataset that both ESG and financial parameters of publicly listed companies worldwide were combined. They presented a framework for conducting statistical analysis and leveraging machine learning techniques to importance of ESG parameters for investment decisions. Their results indicated that the companies with the best ESG ratings had a higher 'return on equity' compared to rest of the companies by highlighting the benefits of financially strong ESG practices.

Simone perazzoli et al. [5] evaluated ESG from a systemic perspective utilizing various criteria. Environment criteria indicate increasing initiatives linked to energy management and emissions, however the major challenges such as the lack of investments, regulation and inspection especially in developing economies remain unsolved. In Social criteria reveal issues concerning employee health, labor practices and employee health safety which may result in violations of fundamental human rights, freedom of association and a safe physical and mental environment. Governance criteria indicated downfall in business ethics pointing to problems with respect to moral norms, accountability, transparency, justice and rights among others. By adopting the General System Theory, they enabled to assess in detail a real case of ESG-related problem.

Hangju et al. [6] stated that rapidly changing global economy and sustainable development of society require practice of ESG management. The model focuses on collected news data on ESG management and analyzed keywords using Big data Analysis. The outcome is to suggest strategic directions for successful ESG management of companies for any changes in the business environment.

Min Gyeong Kim et al. [7] stated that ESG risk ratings are becoming increasingly important to investors, who often use them as a key factor in deciding which companies or industries to invest in. However, there is a lack of research on the relationship between companies' financial disclosures and their ESG risk ratings. To address this gap, they analyzed the risk factors of S and P 500 companies between 2016 and 2020 using BERTopic for topic modeling. In results show that text-based topics reduced the accuracy of predicting ESG risk ratings during the sample period. However, the author observed an improvement in the model's performance after 2019 when a smaller number of topics were extracted. These findings underscore the importance of ESG risk ratings for both investors and businesses.

Pawel A. Lontsikh [8] presents an approach to tackle the issue of sustainable development by implementing ESG criteria. The strategy is based on utilizing robust parametric design methods to implement an integrated environmental process and solve management tasks amidst uncertain conditions. The integrated architecture process follows the ISO/IEC/IEEE 42010 standard and includes documentation and certification processes. To comply with ESG criteria information security and sustainable development standards and the Recommended Practice for Software Requirements specifications must be considered. They analyze the degree of compliance of tasks with forest management standards FSC.

Stefan Pasch, Daniel Ehnes [9] proposed fine-tune transformer-based model for ESG domain. In this model created by combining ESG rating with text documents from annual reports and train as ESG sentiment that predicting the ESG behavior of companies. Authors used NLP to classify the text from annual reports.

3 Architecture Framework

Organizations collect enormous volumes of data on environmental factors. Framework help make ESG assessments more thorough. This data can be processed by artificial intelligence algorithms, which can then extract insightful information and spot trends to assist firms in making decision to reduce ESG rating. Companies can pinpoint areas for development, monitor, and precisely gauge the results of their sustainability programmes, when they have access to real-time information. Organization produce comprehensive and consistent reports by streamlining their ESG reporting procedures with the use of framework. The proposed framework with technology integration is shown in Fig. 1.

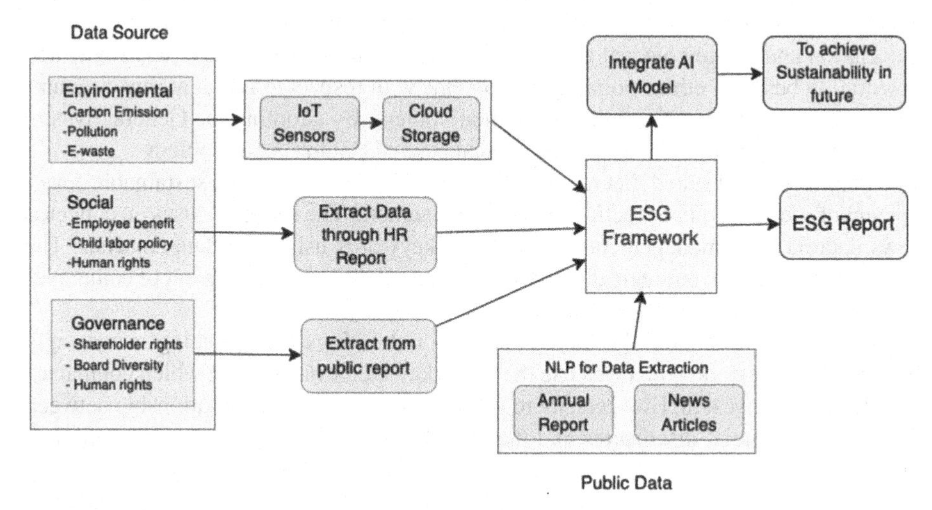

Fig. 1. ESG Framework

3.1 IoT Sensors for Environment Data Collection

The architecture of the Internet of Things (IoT) is essential for gathering data in real-time on a variety of environmental factors. IoT uses three layer for data transfer from sensors or devices, The three layer are perception, network, and application layers. Data acquisition is facilitated by wireless sensors equipped with processing units. These sensors track variables like water waste, CO2 emissions, energy consumption and air quality. They are integrated into various operating settings. Numerous sensors, including those for oxygen, catalytic gas, PM2.5, PM10, ozone (O3), carbon monoxide (CO), carbon dioxide (CO2), nitrogen dioxide (NO2), formaldehyde (HCHO), total volatile organic compounds (TVOCs), Sulphur dioxide (SO2), noise, humidity, and ultraviolet radiation, measure the carbon footprint and the quality of the air both indoors and outdoors. When combined with edge computing units, these sensors allow data to be processed and sent to central systems instantly. In a manufacturing plant, IoT sensors and machine learning algorithms optimized energy consumption by making real-time adjustments to processes, resulting in a significant decrease in the company's carbon footprint.

3.1.1 AI for Social Data Collection

Artificial intelligence is thus useful when collecting and analyzing the social data that would be used in ESG frameworks. Among the most useful techniques for extracting information from numerous textual sources are those of natural language processing, including news stories, annual reports, and social media. Sentiment analysis allows organizations to evaluate the way stakeholders and the general public perceive the social undertakings of companies. AI-based technologies are able to analyze information on labor practices, occupational safety, workforce composition, and involvement of stakeholders. In this way, the organization can trace its social impact with real-time information that assures conformity with social norms. Further, beyond discovery of patterns and trends in sociological data, machine learning algorithms for CSR also help in unveiling much deeper insights into issues like inequality and human rights violations.

3.1.2 AI for Governance Data Collection

Governance data is key to ESG reporting in the organization's structure. AI can be employed in the monitoring of corporate governance structures, the remuneration of CEOs, the diversity of the board, and the rights of shareholders. The use of machine learning algorithms allows for the processing of vast amounts of governance data for the identification of anomalies and potential concerns regarding corporate governance. AI-powered compliance monitoring systems can be used to check internal policies and regulatory updates at any given moment in order to ensure adherence to governance guidelines. Through the use of these tools, it would prove their commitment to moral conduct and strong management systems in the view of the stakeholders by enabling them to provide accurate and comprehensive governance reports. It is their alignment with ESG that underlines the positive change the efficient framework would bring, and organizations can consistently meet stakeholder expectations and regulatory demands by optimizing ESG performance with the help of AI. In addition, it will improve cooperation and trust among stakeholders while enhancing accountability and transparency.

3.2 Natural Language Processing

NLP are necessary for extracting information from different text sources like news articles | annual reports etc. A company can be ranked on social performance using this approach and compared with the social standards in ESG. NLP plays a major role in many analytical activities such as sentiment analysis, public perception interpretation and event extraction keyword-based automated summary. Through NLP and speeding-up social reporting data extraction for organisations to gain critical learnings from their social initiatives/practices, thereby making more informed decisions on sustainability & ethical business practices [10].

3.2.1 Sentiment Analysis

Sentiment is the process of determining if a piece of writing has reflections emotional overtones. This helps us to easily identify whether the opinions of stakeholders shared across various documents and media are impartial, negative or positive. Organizations can potentially use sentiment analysis for accessing the feedback or perception of its stakeholder and public towards ESG initiatives. So they can understand and improve their social projects (and implementations of policies.

3.2.2 Text Classification and Categorization

Text classification is the process of automating a large amount of text documents and assigning them pre-defined categories. Unstructured data is a significant component of ESG reports. NLP facilitates in organizing the texts by topics (like social responsibility, environmental impact or governance issues), allowing a more structured overview on different ESG performance metrics.

3.2.3 Entity Recognition and Event Extraction

NER refers to the process of identifying and classifying text entities such as individuals, companies, locations/dates/technical terms. Event Extraction: Extracts important events in the text. NER can be used to extract entities like cropped image or logo of project, company name and location from the news articles and ESG reports. Event Extraction enables organizations to rapidly respond and remediate on critical events, such as regulation changes or violations (e. g., environmental).

3.2.4 Keyword and Topic Extraction

Topic modeling is methods for extracting the default abstract topics in a set of documents while keyword extractions are used to idenfied significant words or phrases. Keywords and topics extracted from ESG reports, and related documents may help discover themes or focus areas for organizations. It helps understanding of the major topics that are currently being covered with ESG, and drives reporting as well as strategy decision-making.

3.2.5 Automated Reporting and Summarization

Shortening texts into digestible bits and at the same time keeping all necessary information, naturally involves summarization techniques. Automated summarization tools quickly produce summarized SMS messages that distill complex or long ESG reports for easy consumption by stakeholders. This is very useful, in particular for board members as well as investors or regulatory authorities that need to process large quantities of information fast.

3.3 AI in Enhancing ESG Practices

Artificial Intelligence empowers ESG practices with the cutting-edge, AI-enabled framework on social responsibility, the environmental impact, and governance. For instance, predictive analytics powered by AI helps organizations forecast market trends, developing long-term plans that align with sustainable operations. Risk assessment and compliance monitoring can be performed using AI. With continuous analysis of data from different sources, artificial intelligence is able to detect possible compliance issues before they grow into critical problems. It also helps to track sourcing and movement of the material to make sure its constitutions are adherent to both ethical and sustainable standards, therefore improved transparency in the supply chain. For instance, it also can do detailed big data analytics of consumption patterns to provide advice on improvements; hence, the organization will be in a better position to optimize resources and improve energy efficiencies. For example, AI optimizes manufacturing processes so that waste and energy usage can be minimized, hence contributing positively toward an organization's ethos of being sensitive to the environment.

It would automate the collection of data and its analytic processes, thus cutting down the complex parts in ESG reporting. It increases the credibility and precision of ESG with its automated data and decreases the time & effort associated with reporting. Through AI and ESG, companies are given an answer that will serve to break through some of the slimmest odds in corporate history while supporting a more conscientious business landscape. Utilization of AI may enable organizations to live up to stakeholders' expectations and regulatory requirements regarding their ESG performance.

3.3.1 AI Models and Techniques Used in ESG

a) Predictive Analytics: AI models, including time series analysis and regression models, determine impacts on the environment based on historical and present data patterns. Using such projections, businesses can plan how to use them in waste reduction, optimizing their energy use, and shrinking their environmental footprints.
b) Risk assessment: Bayesian networks and decision trees for ESG-related risk assessment. These models help in identifying risks and the estimation of the probabilities of some of the risks like changes in regulations, ecological happenings, among others, which are very important inputs toward their mitigation.
c) Compliance Monitoring: To determine the compliance level with the set social and environmental standards, text analysis from news articles, internal reports, and other regulatory documents is made possible by the merging of Natural Language Processing with rule-based systems. These systems further provide automated metrics tracking for compliance and can flag off deviations early enough for immediate rectification.
d) Supply Chain Transparency: Artificial intelligence used in conjunction with blockchain technology helps supply chains better trace and be more transparent, like tracing materials or goods movement from their sources up to the final consumer, ensuring that they have met the criteria of being ethical and sustainable.
e) Resource optimization and energy efficiency: Optimization algorithms and reinforcement learning analyze the consumption patterns and suggest improvements. These models reduce waste and energy usage by optimizing production and maximizing energy utilization, thus directly contributing to the environmental goals of any organization.
f) Automated ESG Reporting: Through RPA and NLP, the technology automatically collects, evaluates, and integrates data from multiple sources to facilitate ESG reporting. Such technological platforms ensure the accuracy and integrity of reporting with less time and effort spent to prepare in-depth ESG documentation.

3.4 Data Visualization in ESG Reporting

Effective ESG score presentation can be achieved through a clear and legible format of data visualization. Such types of dashboards, charts or graphs are some of the data visualizations that the system uses to represent complex ESG information. These visual aids serve to enhance better understanding and interpretation by providing all those interested with an in-depth view of how a business is performing against ESG criteria. The use of data visualization tools also enhances transparency when judging the ethical behavior of firms because they make it easy for people to access and understand data.

By way of interactive dash-boards users can explore different facets of ESG performance including social initiatives, governance structures, and environmental impact. This ensures that specific areas of interest can be investigated exhaustively to facilitate well-thought-out decision-making. ESG Reporting Framework uses Artificial Intelligence (AI), Machine Learning (ML) and Natural Language Processing (NLP) techniques to transform vast amounts of textual information into actionable insights. These technologies extract pertinent information from things like news articles, reports and other sources through processing and analyzing large volumes unstructured text. When one employs visuals like heat maps; trends become easier to identify which help determine the advantages or weaknesses regarding corporate ESG practices.

These tools also make it easier for organization to benchmark against competitors for follow ESG practices. This strategy fosters better stakeholder communication while also encouraging ongoing ESG practice improvement, which increases corporate accountability and sustainability.

3.4.1 Business Intelligence (BI) Tools in ESG Reporting

There is a significant improvement in ESG reporting through data visualization, business intelligence (BI) tools. Platforms like Tableau, Power BI, and QlikView are some of the strong ones available for creating dynamic and interactive visualizations. Organization can use these BI tools to:

1. Integrate Diverse Data Sources: By collecting information from different sources such as company reports, environmental sensors and social media sites, BI tools can provide a holistic view of ESG performance.
2. Design Interactive Dashboards: To make data more engaging and interesting to different stakeholders, users can interact with dash boards that allow them filter or drill down into specific ESG metrics.
3. Benchmark Performance: The use of BI tools helps businesses identify areas for improvement and best practices by comparing an organization's ESG performance against that of competitors or industry standards.
4. Improve Decision-Making: Through aesthetically attractive visuals and easily understandable formats that they take, these reports help in leading decision making on the part of stakeholders regarding various ESG strategies and initiatives.

4 Result and Discussion

Among other challenges, the current ESG evaluation ecosystem is not standardized and subject to opinion-driven methodology – mainly due different metrics across industries, expert judgement scoring with limited verifiability of data. In response, it is suggested that organizations adopt AI and big data to provide consistent evaluations of ESG. It also can help reduce noise on the social and governance side of unstructured data making practical assessments more secure. IoT technologies deployment further assist in the environmental data collection and management with high efficiency but at a cost that has impacts on environment which needs to be addressed before decision making. However, there must be balance; technology needs to move forward but in such a way that it maintains sustainability.

5 Conclusion

The focus of this study is to combine AI with sustainable technology solutions in order to appraise ESG practices with stress on data collection. Among other things, these technological advancements have transformed data gathering by using artificial intelligence, internet of things and natural language processing technologies, resulting in improved efficiency and accuracy levels. It is important for businesses becoming more committed to sustainability that the intersection between technology and ESG extends. When assisted by these technological advancements, ethical and long-term corporate practices can be based on simplification as well as refining information collection. This study contributes to our knowledge of how technology relates with ESG through underscoring the significance of data collecting in upholding moral and sustainable business behavior. The findings further demonstrate the importance of embracing state-of-the-art technological solutions as demand for accountability and transparency heightens in today's business settings.

References

1. Saxena, A., Sharma, B., Verma, P., Aggarwal, R.: Technologies Empowered Environmental, Social, and Governance (ESG): An Industry 4.0 Landscape. Sustainability **15**(1), 309 (2022). https://doi.org/10.3390/su15010309
2. Pozzi, F.A., Dwivedi, D.: ESG and IoT: ensuring sustainability and social responsibility in the digital age. In: Communications in computer and information science, pp. 12–23 (2023). https://doi.org/10.1007/978-3-031-47997-7_2
3. Wu, W., et al.: Consortium blockchain-enabled smart ESG reporting platform with token-based incentives for corporate crowdsensing. Journal in Computers & Industrial Engineering **172**(A) (2022)
4. Gupta, A., Sharma, U., Gupta, S.K.: The role of ESG in sustainable development: an analysis through the lens of machine learning. In: 2021 IEEE International Humanitarian Technology Conference (IHTC), pp. 1–6. IEEE, Toronto (2021)
5. Perazzoli, S., Joshi, A., Ajayan, S., Santana Neto, J.P.: Evaluating Environmental, Social, and Governance (ESG) from a Systemic Perspective: An Analysis Supported by Natural Language Processing. Social Science Research Network Journal (2022)
6. Seo, H., Jo, D. H., Pan, Z.: Big data analysis of ESG news using topic modeling. In: 2022 IEEE International Conference on Big Data (Big Data), pp. 1–6. IEEE, Seattle (2022)
7. Kim, M.G., Kim, K.S., Lee, K.C.: Analyzing the effects of topics underlying companies' financial disclosure about risk factors on prediction of ESG risk rating Emphasis on BERTopic. In: 2022 IEEE International Conference on Big Data (Big Data), pp. 1–6. IEEE, Seattle (2022)

8. Lontsikh, P.A., Golovina, E.Y., Evloeva, M.V., Livshitz, I.I., Koksharov, A.V.: Implementation of ESG Sustainable Development Concept Criteria Using the Robust Design Methods. In: 2022 International Conference on Quality Management, Transport and Information Security, Information Technologies (IT&QM&IS), pp. 1–5. IEEE, Moscow (2022)
9. Pasch, S., Ehnes, D.: NLP for responsible finance: fine-tuning transformer based models for ESG. In: 2022 IEEE International Conference on Big Data (Big Data), pp. 1–6. IEEE, Seattle (2022)
10. Lee, O., Joo, H., Choi, H., Cheon, M.: Proposing an Integrated Approach to Analyzing ESG Data via Machine Learning and Deep Learning Algorithms. Sustainability **14**(14), 5407 (2022). https://doi.org/10.3390/su14145407
11. Dwivedi, D., et al.: A machine learning based approach to identify key drivers for improving corporate's esg ratings. Journal of Law and Sustainable Development **11**(1), e0242 (2023). https://doi.org/10.37497/sdgs.v11i1.242
12. Gregory, R.P.: The influence of firm size on ESG score controlling for ratings agency and industrial sector. Journal of Sustainable Finance & Investment **14**(1), 86–99 (2024)
13. Dwivedi, D., Batra, S., Pathak, Y.K.: A machine learning based approach to identify key drivers for improving corporate's esg ratings. Journal of Law and Sustainable Development **11**(1), e0242 (2023). https://doi.org/10.37497/sdgs.v11i1.242
14. Dwivedi, D.N., Tadoori, G., Batra, S.: Impact of women leadership and ESG ratings and in organizations: a time series segmentation study. Acad. Strateg. Manag. J. **22**(S3), 1–6 (2023)
15. Pozzi, F.A., Dwivedi, D.: ESG and IoT: ensuring sustainability and social responsibility in the digital age. In: Tiwari, S., Ortiz-Rodríguez, F., Mishra, S., Vakaj, E., Kotecha, K. (eds.) Artificial Intelligence: Towards Sustainable Intelligence. AI4S 2023. Communications in Computer and Information Science, vol 1907. Springer, Cham (2023). https://doi.org/10.1007/978-3-031-47997-7_2

Attention-Based Deep Learning for Hand Gesture Recognition Using Multi-sensor Data

Rinki Gupta[1](✉), Ankit Kumar Das[1], and Ghanapriya Singh[2]

[1] Electronics and Communication Engineering Department, Amity University Uttar Pradesh, Noida, India
rgupta3@amity.edu

[2] Department of Electronics and Communication Engineering, National Institute of Technology Kurukshetra, Kurukshetra, Haryana, India

Abstract. Physical activity, such as playing sports, is useful in maintaining a heathy mind-body complex. In this work, multiple wearable surface electromyography (sEMG) and accelerometer sensors have been employed to capture data for hand gestures of cricket umpire. The proposed approach will not only accurately interpret the umpire's decision but may also be utilized during their training. Deep learning models consisting of convolutional layers and recurrent layers are finely tuned to process these different modality data for hand gesture recognition. Moreover, attention mechanism is utilized to enhance the efficacies of the models. The attention scores assist the network in focusing on selective regions of the input feature map, hence improving the performance of the network. The proposed convolutional recurrent neural network (CRNN) along with attention layer achieves an average accuracy of 97.1% for accelerometer data. This is around 2% higher than the accuracy attained by the same model without the attention mechanism. Similar observations are obtained for the sEMG data demonstrating the utility of attention mechanism in time-series data-based classification task.

Keywords: Deep learning · Attention mechanism · Surface electromyography · Accelerometer · Hand gesture recognition

1 Introduction

Engaging in sports not only improves the physical fitness of a person in terms of flexibility, strength and stamina, but also encourages soft skills such as teamwork, discipline and social interaction [1]. In this work, recognition of hand gestures of cricket umpire is considered using data from wearable, non-invasive sensors and deep learning models. Hand gesture recognition finds applications across various fields, including computer vision, robot control, assistive devices, and virtual reality [2]. Its applications span diverse industries and offer a new dimension to human-computer interaction, from controlling applications to enhancing virtual and augmented reality experiences. Deep learning methods, particularly in hand motion recognition, offer promising solutions, such as in cricket refereeing [3]. By analyzing hand movements, deep learning can enhance the accuracy

S. Tiwari et al. (Eds.): AI4S 2024, CCIS 2243, pp. 115–126, 2025.
https://doi.org/10.1007/978-3-031-81369-6_10

of calls and contribute to fair play. Implementing deep learning methods for hand motion recognition of cricket umpire involves leveraging computer vision and image processing techniques to analyze umpire gestures during critical moments. This technology aids the umpires in making accurate decisions while minimizing human error.

Hand gesture recognition systems may utilize sensors such as video cameras, motion sensors, electromyograms, ultrasound sensors, and advanced algorithms to capture and analyze human hand movements. While video sensors are easily available, wearable sensors are unaffected by issues such as background, lighting condition, and obstructed view. For the application considered in this paper, surface electromyography (sEMG) and accelerometer (ACC) sensors are employed due to the aforementioned advantages, which will be useful in an actual playground. The major contributions of this work are summarized as under:

- A database of sEMG and ACC signals is recorded using multiple wearable sensors on hand for 12 activities of cricket umpire with 5 volunteers.
- Convolutional recurrent neural networks (CRNN) with attention are designed for sEMG and ACC data to perform classification of the considered activities accurately.
- The performances of the proposed models are compared with similar models without attention on the cricket umpire dataset.

The proposed work provides a framework for accurately interpreting the umpire's decisions from their hand gestures, which is useful in not only providing quick and accurate decisions but may also aid in developing a training mechanism for novice cricket umpires.

The remaining paper is organized into another 4 sections. Section 2 contains a review of related work on hand gesture recognition and deep learning models for classification. Section 3 presents the details of the collected database and the CRNNs designed for classification of the considered activities. Results of evaluation of models' performances and tuning of the model parameters are presented in Sect. 4. Section 5 concludes the paper along with a brief mention of the scope of future work.

2 Review of Related Literature

In this section, a brief overview of the techniques for capturing hand motions and their processing in deep neural networks for gesture recognition is provided, along with a concise description of the attention mechanism.

2.1 Hand Gesture Recognition

Hand gesture recognition has become a popular topic of research due to the myriad of applications that it is useful for. For instance, human-computer-interfaces, robotic control, prosthetic control, rehabilitation, elderly care, surveillance, and competitive sports [1–5]. Hand gesture recognition has been designed using various sensing techniques, that may be broadly categorized as vision-based and sensor-based techniques. The use of vision-based sensors is prominently reported in literature as pointed out in the review paper [4]. The vision-based setups may employ passive markers that are reflective surfaces, that are placed on the limbs whose motions are to be monitored. On the other hand, with active markers, there may be a source of light, such as light emitting diode or infrared ray emitting diode that are used for tracking the motion of fingers, hands, and even torso, and lower limbs. In vision-based sensors, hand motion may also be captured without markers, using just RGB camera. Authors have also proposed using color gloves while capturing motion and postures of hands, so that it is easier to segment the hand region from the background [6]. A few limitations of using vision-based sensors are that occlusion must be avoided, complex backgrounds may make the task more challenging, and for visible light-based approaches, proper lighting may also affect the data acquisition.

Sensor-based systems for hand gesture recognition have been developed using inertial measurement units (IMU) consisting of accelerometer and gyroscope [4]. While accelerometers provide information about orientation of hand with respect to gravity and its linear acceleration, gyroscopes measure the rate of turn. IMU sensors and video cameras are also found in smartphones, hence, smartphones have been used as an easily accessible alternative to capture hand motion. The challenges while using IMU sensors are that multiple sensors may be required to capture the motion of different segments of hands such as fingers, wrist, forearm, upper arm and shoulders. Also, the placement of the sensor may drastically vary the signals from one user to another. Surface electromyogram (sEMG) sensors that capture electrical potential of muscles are also placed on the surface of the skin on top of various locations of hand muscles to capture the electrical patterns for different types of postures and motions of hands [7, 8]. Other than sEMG, force myography measures motions and position of muscles by recording volumetric changes in muscles and has also been employed for hand gesture recognition [4].

Once the hand gesture data has been captured, it may be processed to prepare the data for generating the required prediction. Data preparation generally includes data cleaning, data visualization, dimensionality reduction, data normalization and data segmentation [2, 8]. Thereafter, machine learning or deep learning models have been employed to perform hand gesture recognition. Several machine learning models including support vector machines, k-nearest neighbors (kNN), random forest, XGBoost [9], and naïve bayes have been popularly used in literature for hand gesture recognition [2, 9]. Since 2020, there has been a significant increase in the use of deep learning models for hand gesture recognition. These include convolutional neural networks (CNN) [10], long short-term memory (LSTM) networks [5], gated recurrent unit (GRU) networks, convolutional recurrent neural networks (CRNN) [2, 8], deep residual networks [7] and transformers [11]. A review of deep learning-based hand gesture recognition for each type of sensing technology is presented in [4].

Hand gesture recognition has been utilized for sports applications as well. For instance, image preprocessing and scale-invariant feature transform (SIFT) are employed to extract image features to develop a system capable of recognizing various umpire signals from images in [12]. The feature dimensions are reduced using k-means clustering and principal component analysis. Subsequently, the signals are classified using three classifiers kNN, decision tree, and random forest with up to 81% accuracy after training on a database of 6000 images across six activity classes. In the paper [13], the authors present a novel method for detecting five key events in cricket videos, including four, six, out, no ball, and wide, by recognizing the umpire's signals by utilizing a pretrained CNN architecture. The framework achieves test accuracies of 97.76% for umpire frame detection and 86.14% for umpire signal recognition by training on a newly introduced dataset, comprising 504 videos and 2000 images from cricket. In [14], the authors analyzed video frames of the umpire's stance and classify the common event classifications like six, no ball, wide, out, leg bye, and four. They employed CNNs to attain 98.20% accuracy on a dataset of 1040 Umpire Action Images showcasing these events. In [15], the authors employed sEMG and IMU signals for transfer learning and reached high average accuracy with deep belief networks for recognition of official referee's signals. We particularly focus on the use of attention mechanism for hand gesture recognition, a brief description of which is presented in the next sub-section.

2.2 The Attention Mechanism

Attention mechanism opens a new dimension to learning from data using deep neural networks. The attention mechanism is inspired by how humans selectively process the most relevant information while understanding their environment. With attention, the network learns to focus on the most relevant features extracted from the input data, improving its efficiency and performance. This is done by learning attention score for a given input feature map, that would weigh the input features according to relevance. Although, the history of attention mechanism dates back to late 1980's, the rise in its usage was observed after mid 2010's [16]. Attention mechanisms may be broadly categorized based on softness of attention, forms of input features, and input and output representations as reviewed in [17]. In distinctive attention, the query and the key that are used for calculating the attention scores are derived from two separate sequences, while in self-attention, the query (Q), the key (K) of dimension d_k and the value (V) are derived from the same input sequence. The scaled dot-product attention function is evaluated as [18]

$$\mathrm{Att}(Q, K, V) = \mathrm{SoftMax}\left(\frac{QK^T}{\sqrt{d_k}}\right)V. \tag{1}$$

When Q and K are derived from the same sequence, the attention function is calculated for every segment of the sequence against all the other segments in the sequence. This defines the importance of the values of each segment relative to the others.

Self-Attention mechanism has been employed with image data as well as sensor data for human activity recognition. In [11], for instance, the authors employ self-attention in an action transformer that takes a sequence of images as inputs to determine human pose. A multi-scale attentive feature fusion block is presented in [19] to extract robust spatial features for hand gesture recognition with a lightweight CNN. The authors of [20] designed a CNN model with attention modules to perform human activity recognition from weakly labelled sensor data. The attention modules helps the model to focus on most relevant regions of the stream of sensor data improving its performance in human activity recognition. In [21], sEMG data recorded from the myoarmband is processed using a CNN with attention blocks to classify 12 hand gestures with 91.64% accuracy, which was around 1% higher than that attained with a CNN model with no attention blocks. In this work, CRNN models with attention are proposed for classification of hand gestures of cricket umpire, as discussed in the next section.

3 Proposed Methodology

This section contains the details of the multi-sensor database collected for hand gestures of cricket umpire and the deep learning models proposed for their classification.

3.1 Database Details

The database was recorded with 5 male volunteers, each performing 12 hand gestures associated with cricket umpires, examples of which are shown in Fig. 1. These activities corresponded to cricket umpire's gestures for Out, Not Out, No Ball, Free Hit, Wide Ball, Four Runs, Six Runs, Bye, Bouncer, DRS, Revoke Decision, and Penalty Runs, numbered as activity 1 to 12 during classification, in the same order. Each volunteer performed every gesture ten times, resulting in a total of 600 observations in the database. An audio stimulus was played to guide the volunteer when to start performing the activity.

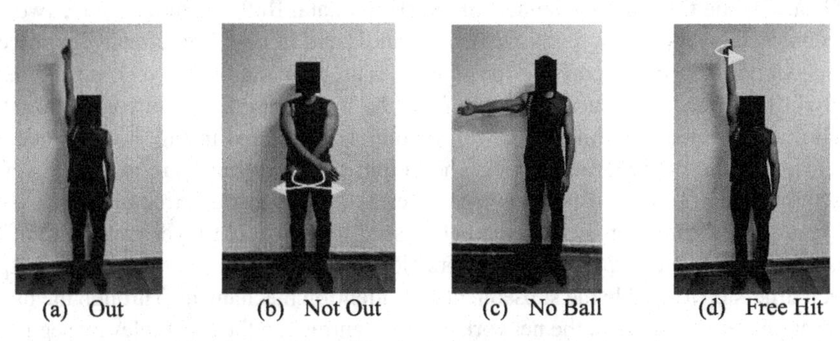

(a) Out (b) Not Out (c) No Ball (d) Free Hit

Fig. 1. Hand activities in the cricket umpire dataset.

The database was recorded using the EMG + IMU system from Delsys Inc., shown in Fig. 2a. Each Avanti sensor captures single channel sEMG and three-axis ACC data at 1259 Hz and 148 Hz, respectively. The sensor data is digital at 16-bit resolution and is wirelessly transmitted to the EMG + IMU system, from where it is sent to the computer for live display and storage. Eight Avanti sensors were placed on the subjects, 4 on each hand, as shown in Fig. 2b. The sensors were located on major muscles of the hand to capture useful sEMG data, while the hand motion is also recorded with the ACC sensors.

The recorded sEMG and ACC data was then prepared so that it may be given to the deep learning model as inputs. The non-zero mean of sEMG signals were subtracted for baseline removal. The average of absolute values of the ACC signals recorded for the sensors on the left arm and the right arm were determined. Whenever the average of the ACC exceeded a certain threshold, the activity region was detected. A 5s segment of the signals starting from the beginning of each activity region were separated and provided as independent observations in the dataset. The data was processed using the CRNN model described in the next sub-section.

(a) EMG+IMU system

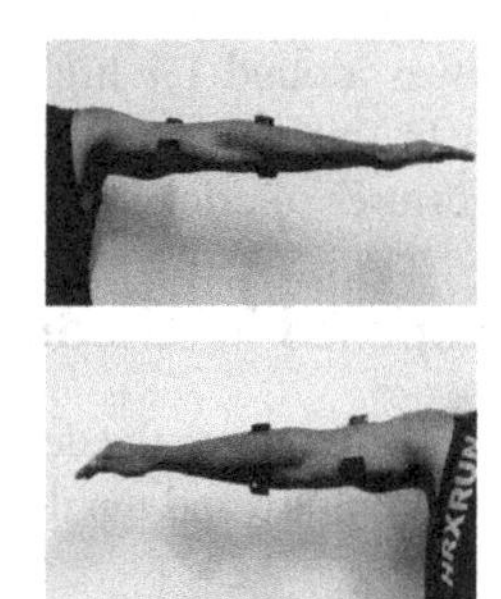

(b) Sensor Placement

Fig. 2. Data acquisition setup.

3.2 Proposed Deep-Learning Models

The architecture of the CRNN model proposed for classification of hand gestures from the cricket umpire database is shown in Fig. 3. Similar model architectures are employed for both ACC and sEMG data. Since 8 Avanti sensors are used, there are 24 channels in ACC data, while there are 8 channels in the sEMG data. Both the models have two 1D convolutional layers with hyperparameters as indicated in Fig. 3, to effectively process the sequential data obtained from the sensors. The convolutional layers are the main highlight for extracting generic features from the input data. Each convolutional layer is followed by a max-pooling layer with a pool size as stated in Fig. 3, to reduce the dimension of the output feature map. The output of the second max-pooling layer is passed through a TimeDistributed layer to preserve the time-step structure of the data. The resulting feature map is flattened and passed to the LSTM layer consisting of 64 units. The LSTM layer captures the temporal dependencies in the data and returns output at each time-step to enable the subsequent self-attention mechanism. Through the usage of the attention mechanism, the network can concentrate on the most relevant segments of the input sequence, which in turn enhances the model's capability to differentiate between various gestures in a more precise manner. Dropout with rate 0.5 is introduced to avoid overfitting in the model. The output of the attention layer is flattened and passed to the final dense layer with SoftMax activation and 12 units that provide the prediction of the final class probabilities. The shape of the output feature maps for each layer in the CRNN model used for processing the ACC and the sEMG data, and the number of trainable parameters is listed in Table 1. The ACC model has fewer trainable parameters due to less dimension of input because ACC is sampled at a lower rate than sEMG.

Substantial hyperparameter optimization is performed with the aim of improving the performance of the proposed models. This includes changing such factors as the number of convolutional and LSTM layers, the number of filters in the convolutional layers, pool size in max-pool layer and units in the LSTM layer. The tuning of the model allows the models to be fit for purpose with specifications of the sEMG and accelerometer data making hand gesture recognition reliable and accurate.

For training, 80% of the dataset was provided to the model, while the remaining 20% was used for testing. The trainable parameters of the network were optimized by minimizing the sparse cross-entropy loss function Adam optimizer with a learning

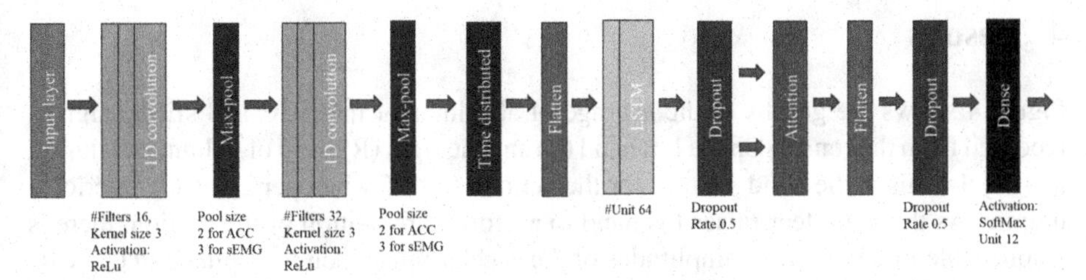

Fig. 3. Proposed CRNN model for classification of cricket umpire hand gestures.

Table 1. Summary of proposed CRNN model.

No	Layer	For ACC data		For sEMG data	
		Output Shape	Parameters	Output Shape	Parameters
0	Input	815, 24	0	6926, 8	0
1	Convolution 1D	811, 8	968	6924, 16	400
2	Max-pool	405, 8	0	2308, 16	0
3	Convolution 1D	401,16	656	2306, 32	1568
4	Max-pool	200, 16	0	768, 32	0
5	Time Distributed	200, 16	0	768, 32	0
6	Reshape	200,16	0	768, 32	0
7	LSTM	200, 64	20736	768, 64	24832
8	Dropout	200, 64	0	768, 64	0
9	Attention	200, 64	0	768, 64	0
10	Flatten	12800	0	49152	0
11	Dropout	12800	0	49152	0
12	Dense	12	153612	12	589836
	Total Parameters (All trainable)		175972		616636

rate of 0.001 for ACC data and 0.0007 for sEMG data was used for effective learning, batch size was set to 16 and the highest number of epochs was set to 100. The training process comprises of early stopping with a patience of 5 epochs on the validation loss to avoid overfitting. The model is assessed based on the classification accuracy and loss of the model on the test dataset. The models have been implemented in Python using TensorFlow and Keras libraries. The results for evaluation of the above model on the cricket umpire database are given in the next section.

4 Results

Figure 4 shows the graphs of the average of absolutes of the ACC and sEMG signals recorded from the sensors on the left hand (L) and the right (R) hand of volunteer 1 during activity 1. This is the hand gesture for the action of OUT when performed by a cricket umpire. As the volunteer raises his hand to perform a repetition of the motion, there is a noticeable increase in the amplitudes of the accelerometers on the right hand (ACCR) and electromyograms on the right hand (EMGR), indicating heightened activity in the muscles and movement of the right hand. Conversely, the readings for the accelerometers on the left hand (ACCL) and electromyograms on the left hand (EMGL) remain relatively stable or show minimal changes. This suggests that the raising action primarily involves the right hand, with the left hand potentially providing support or stability rather than generating significant force. The detected activity regions are indicated by a value 1, while rest is indicated by a 0 in the graphs. The data for each detected activity region is considered as a separate observation in the database. Hence, the signals shown in Fig. 4 yield 10 observations in the database.

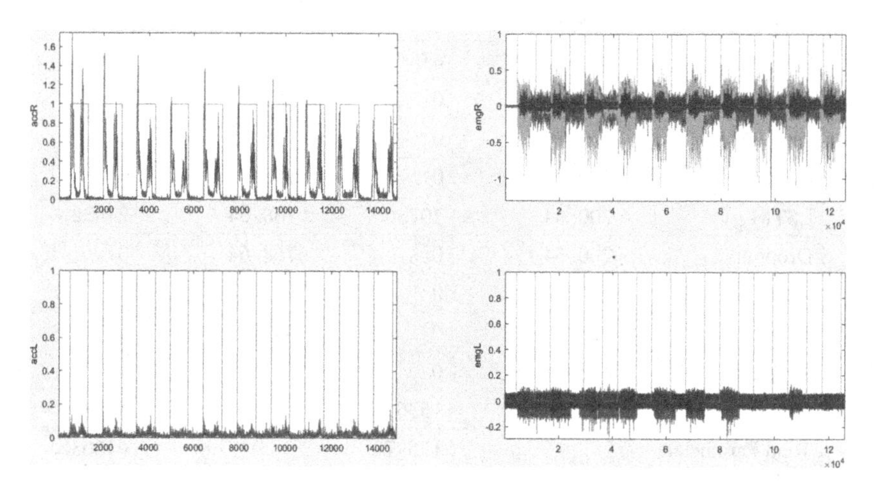

Fig. 4. Activity detection for signal segmentation.

The ACC and sEMG data are then processed in the CRNN model, taking model parameters as indicated in Fig. 3. The accuracy vs. epoch and the loss vs. epoch for the ACC model and the sEMG model, with attention are shown in Figs. 5a and b, respectively. It may be observed that both the models train without overfitting. Overall, the loss for both the models decreases and accuracies increase with epochs. The models generalize well yielding similar results on test data. The sEMG model took more epochs to train since the number of samples was larger and the learning rate was lowered to achieve smoother convergence.

The accuracies achieved for each of the considered cricket umpire gestures using the ACC and sEMG models with attention are plotted in Fig. 6. For the considered data, better accuracies are attained with ACC data. This is mainly because each activity consists of some motion of hands that leave a distinct pattern in the ACC data. On the other hand, sEMG data is stochastic in nature, and more challenging to analyze with deep learning models.

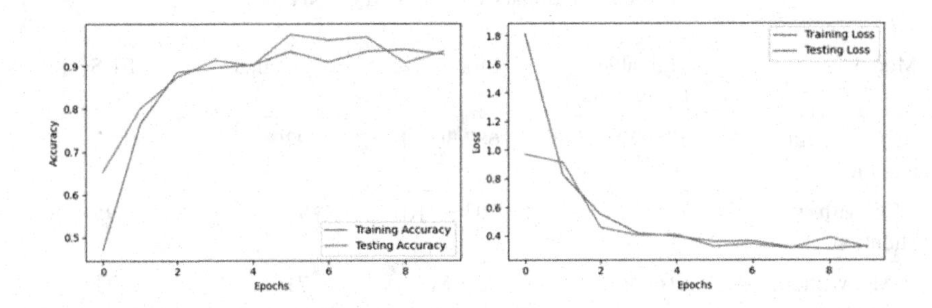

(a) Accuracy vs. epoch and the loss vs. epoch for the ACC model with attention

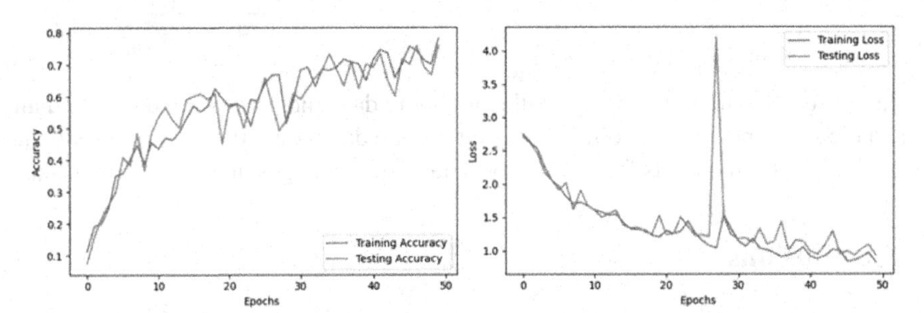

(b) Accuracy vs. epoch and the loss vs. epoch for the sEMG model with attention

Fig. 5. Learning graphs for ACC and sEMG models with attention.

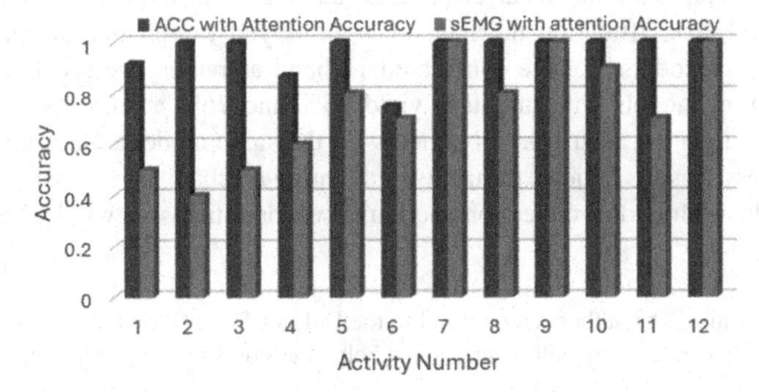

Fig. 6. Class-wise accuracies for ACC and sEMG models with attention.

After early stopping, the best performing models gave the average accuracies over all activities as stated in Table 2. With attention, the accuracies for ACC and the sEMG signals are better as compared to that obtained with the same model architectures, with no attention layer. This indicates the utility of attention mechanism in processing time-series data for the classification task.

Table 2. Summary of proposed CRNN model.

Model	Trainable Parameters	Model size	Accuracy	F1 Score
ACC without attention	175972	687.39 KB	92%	0.92
ACC with attention	175972	703.34 KB	95%	0.95
sEMG without attention	616636	2.35 MB	71%	0.71
sEMG with attention	616636	2.38 MB	76%	0.76

In future, we will work towards collecting more data and processing multi-modality data in the same model. The collected cricket umpire database is one of its kind and may be used to pre-train models for any sensor-data based hand gesture recognition task.

5 Conclusions

In this paper, classification of hand gestures of cricket umpires is considered using wearable sensor data. A database of accelerometers and electromyograms is collected from multiple volunteers using multiple sensors placed on both hands. The collected data is pre-processed and given as input to convolutional recurrent neural network. The proposed convolutional recurrent neural networks are tuned to perform well on the considered task. Attention mechanism is employed to enhance the performance of the proposed models. For the considered 12 hand activities, the accelerometer and electromyogram models with attention yield 95% and 76% accuracies, respectively. This is higher than the accuracies obtained with the same model architectures without attention layers. Results indicate the utility of attention mechanism in classification using time-series data. More data collection and multi-sensor data fusion will be performed in future.

Acknowledgments. The authors are grateful to the DeLuca Foundation for the individual grant and for providing the data acquisition device that helped us collect the data reported in this work.

Disclosure of Interests. The authors have no competing interests to declare that are relevant to the content of this article.

References

1. Misha, K., Khalid, S., Aleryani, A., Khan, J., Ullah, I., Ali, Z.: Human action recognition systems: A review of the trends and state-of-the-art. IEEE Access (2024)
2. Pyun, K.R., et al.: Machine-learned wearable sensors for real-time hand-motion recognition: toward practical applications. Natl. Sci. Rev. **11**(2), 298 (2024)
3. Sharma, D.: Review of Application of Gesture and Poses for Reducing Injury in Sports. Int. J. Converg. Healthc. **4**(1), 31–35 (2024)
4. Noh, D., Yoon, H., Lee, D.: A decade of progress in human motion recognition: a comprehensive survey from 2010 to 2020. IEEE Access (2024)
5. Gupta, R., Madan, P., Srinivasan, G.: Deep learning models for video-based hand gesture recognition in robotic applications. 9th International Conference on Advanced Computing and Communication Systems, IEEE, 1, pp. 2168–2172 (2023)
6. Guo, L., Lu, Z., Yao, L.: Human-machine interaction sensing technology based on hand gesture recognition: a review. IEEE Trans. Human-Machine Sys. **51**(4), 300–309 (2021)
7. Khattak, A.S., et al.: Hand gesture recognition with deep residual network using sEMG signal. Biomedical Engineering/Biomedizinische Technik **69**(3), 275–291 (2024)
8. López, L.I.B., Ferri, F.M., Zea, J., Caraguay, Á.L.V., Benalcázar, M.E.: CNN-LSTM and post-processing for EMG-based hand gesture recognition. Intel. Sys. Appl. **22**, 200352 (2024)
9. Gupta, R., Bhatnagar, A.S.: Multi-stage Indian sign language classification with Sensor Modality Assessment. IEEE 7th International Conference on Advanced Computing and Communication Systems, pp. 18–22 (2021)
10. Gupta, R., Bhatnagar, A.S., Singh, G.: A Weighted Deep Ensemble for Indian Sign Language Recognition. IETE Journal of Research, 1–8 (2023)
11. Mazzia, V., Angarano, S., Salvetti, F., Angelini, F., Chiaberge, M.: Action transformer: A self-attention model for short-time pose-based human action recognition. Pattern Recogn. **124**, 108487 (2022)
12. Wyawahare, M., Dhanawade, A., Dharyekar, S., Dhole, A., Dhopade, M.: Automating Scorecard and Commentary Based on Umpire Gesture Recognition. Advancements in Smart Computing and Information Security, Springer, **1759** (2022)
13. Sakib, S., Mridha, M.A.H., Hasan, N., Habib, M.T., Mahbub, M.S.: Event Detection from Cricket Videos Using Video-Based CNN Classification of Umpire Signals. 4th International Conference on Sustainable Technologies for Industry 4.0, 1–5 (2022)
14. Nandyal, S., Kattimani, S.L., Halakatti, P.G.: Cricket event recognition and classification from umpire action gestures using convolutional neural network. Int. J. Adv. Comput. Sci. Appl. **13**(6) (2022)
15. Pan, T.Y., Tsai, W.L., Chang, C.Y., Yeh, C.W., Hu, M.C.: A hierarchical hand gesture recognition framework for sports referee training-based EMG and accelerometer sensors. IEEE Transactions on cybernetics **52**(5), 3172–3183 (2020)
16. Soydaner, D.: Attention mechanism in neural networks: where it comes and where it goes. Neural Comput. Appl. **34**(16), 13371–13385 (2022)
17. Niu, Z., Zhong, G., Yu, H.: A review on the attention mechanism of deep learning. Neurocomputing **452**, 48–62 (2021)
18. Vaswani, A., et al.: Attention is all you need. Advances in neural information processing systems 30 (2017)
19. Bhaumik, G., Govil, M.C.: SpAtNet: A spatial feature attention network for hand gesture recognition. Multimedia Tools and Applications **83**(14), 41805–41822 (2024)

20. Wang, K., He, J., Zhang, L.: Attention-based convolutional neural network for weakly labeled human activities' recognition with wearable sensors. IEEE Sens. J. **19**(17), 7598–7604 (2019)
21. Zhang, Z., Shen, Q., Wang, Y.: Electromyographic hand gesture recognition using convolutional neural network with multi-attention. Biomed. Signal Process. Control **91**, 105935 (2024)

Climate Risk Management for Aquaculture Industry: Robust Programming Approach vs Random Forest Algorithm

Beren Gürsoy Yılmaz[1,2](✉) and Ömer Faruk Yılmaz[1,2]

[1] Department of Industrial Engineering, Karadeniz Technical University, 61080 Trabzon, Turkey
berengursoy@ktu.edu.tr
[2] Department of Industrial and Systems Engineering, University of Florida, Gainesville 32608, USA

Abstract. This paper addresses the problem of fish escapes in trout farming within the context of climate risk management for the aquaculture industry, with the objective of cost minimization for Kaizen projects. To address this problem, two different approaches are proposed: (I) a robust optimization-based approach and (II) a random forest-based approach. The robust approach aims to minimize cost by focusing on the worst-case scenario, while the random forest approach predicts the cost associated with the number of escaped fish by considering the levels (high and low) of features affecting the response variable. A case study from an enterprise in the Black Sea region of Turkey, a significant exporter of seafood, is presented to illustrate the problem and its external risk causes. The effectiveness of both approaches is evaluated through multiple instances. The results reveal that the random forest-based approach should be preferred when 60–70% of the features have low levels. Conversely, the robust approach should be preferred when most of the features have high levels to reduce the cost incurred by group-based Kaizen ap-plications. The findings highlight the potential of these methodologies to enhance sustainability in the aquaculture industry. Future research directions include extending the problem to cover different stages of the aquaculture supply chain and adopting other machine learning algorithms to improve system performance.

Keywords: Climate risk management · Aquaculture industry · Fish escaping problem · Robust programming approach · Random forest algorithm

1 Introduction

Aquaculture, which encompasses a wide variety of freshwater and marine fish, shellfish, crustaceans, and aquatic plants, has become one of the most essential primary industries and a key driver of economic activity in both local and global economies [1].

With the development of aquaculture, concerns about the external risks related problems have arisen in the industry. Therefore, there is a need to gain further knowledge on

S. Tiwari et al. (Eds.): AI4S 2024, CCIS 2243, pp. 127–140, 2025.
https://doi.org/10.1007/978-3-031-81369-6_11

this issue and provide comprehensive solutions. In this manner, farmed fish escape is a major concern in the aquaculture industry within the context of external risks, and there is limited understanding of the accident scenarios involving fish escape [2]. Hence, the systematic analysis of information and data about these types of incidents may reveal contributing causes [3].

Motivated by this issue, the present study examines the fish escape problem through a case study of an enterprise conducting fish farming operations in the Black Sea region of Turkey. Although focusing on the Black Sea region may limit the generalizability of the results, this area is specifically chosen for its favorable climate conditions such as the ideal water temperature and salt levels for rainbow trout farming. As a result, the Black Sea has become one of the world's largest exporters of seafood, with the trout farming industry being a significant source of revenue [4, 5]. However, despite its great potential for further growth, sustainability challenges related to escaped fish have hindered the industry's development. These considerations are critical to the study's objectives and provided a relevant context for analyzing the phenomena under investigation.

Fish farming companies face significant financial losses from large-scale fish escapes due to lost biomass [6]. Moreover, these events can inflict severe reputational damage, which is considered the highest strategic risk for both the companies and the industry [7]. Comparing fish farming accidents to major accidents in other industries reveals that large-scale fish escapes should be regarded as major accidents because of their severe environmental impact [8].

Therefore, there is a need to implement a lean-oriented methodology to mitigate the severe consequences of the fish escape problem. Since one of the main techniques in lean applications is group-based Kaizen projects, this study considers them to be the primary precaution to reduce the number of escaped fish. Although group-based long-term Kaizen activities have proven to be highly effective in addressing problems related to external and operational risks [9], they also incur cost for conducting all necessary operations as countermeasures to overcome the root causes of the main problem.

In this study, two different approaches are applied to reduce the cost of Kaizen projects: (I) a robust optimization-based approach and (II) a random forest-based approach. The first approach aims to minimize Kaizen-related cost by focusing on the worst-case scenario, while the second approach employs a random forest regressor to predict the cost associated with the number of escaped fish by considering the levels (high and low) of features affecting the response variable.

The rest of the study is organized as follows: Sect. 2 is devoted to the problem description along with the features affecting the fish escape problem. Section 3 introduces a case study and the applied approaches. Section 4 presents the computational analysis. Finally, Sect. 5 provides conclusions along with findings and insights.

2 Problem Description

In this study, climate risk management for the trout farming industry within the context of aquaculture is investigated. After a careful investigation and long-term research with industry stakeholders, it has realized that the main problem in the industry is trout escaping from the farms, which is visualized in Fig. 1.

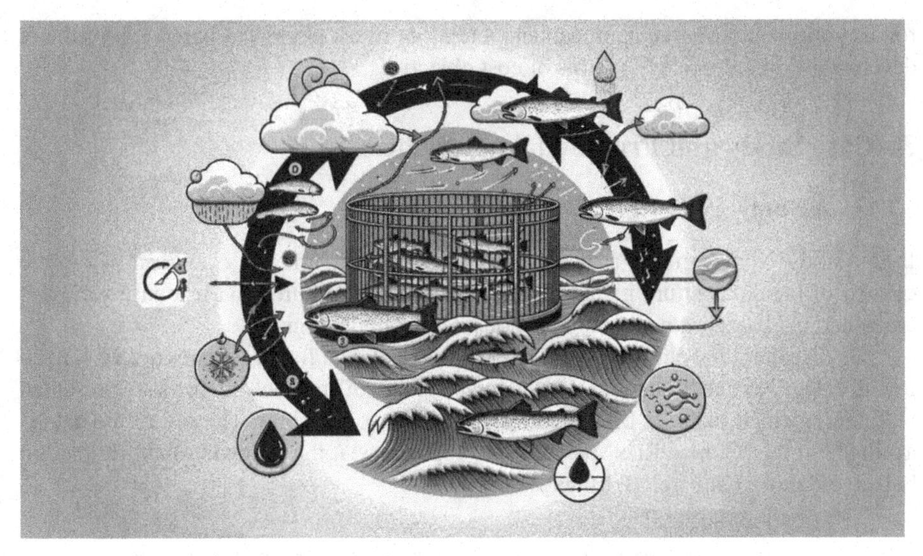

Fig. 1. Illustration of trout escapes from nets

According to experts' opinions, there are five main factors that cause fish escapes from the farms, which can be classified under the category of external risks. The risks can be examined either in external or internal/operational risk categories [10].

Currents Speed (CS): Strong underwater currents (measured in meters per second, m/s) can strain and tear the nets. The direction and speed of the currents directly affect the durability of the nets.

Rainfall Intensity (RI): Heavy rainfall (measured in millimeters per hour, mm/h) can change the salinity and density of the water, causing stress on the nets and their connections.

Wind Speed (WS): Strong winds (measured in kilometers per hour, km/h) can create large waves on the sea surface, leading to net tears. The direction and speed of the wind, along with wave height and energy, challenge the durability of the nets.

Waves Height (WH): Storms and large waves (wave height measured in meters, m) can directly impact the nets, causing them to tear. The duration and intensity of the storms increase the stress on the nets.

Salinity Changes (SC): Sudden changes in salinity (measured in parts per thousand, ppt) can affect the durability of the net materials. Particularly after rain, the accumulation of freshwater on the sea surface can cause such changes.

In order to mitigate the negative impacts of both risk categories, enterprises can implement Kaizen projects [11]. When addressing operational risks, individual Kaizen initiatives can be implemented before and after specific tasks, and they can be completed successfully. However, for external risks, group-based Kobetsu Kaizens can be preferred to achieve long-term and effective solutions, even though they may incur higher cost [12].

In this study, group-based Kaizen projects are considered to address the problem of trout escapes since it is categorized under the external risk category. To this end, both a robust optimization-based approach and a random forest algorithm-based approach are developed for comparison in terms of cost objectives.

3 Case Study and Proposed Approaches

3.1 Case Study

In this study, trout farming in the Black Sea Region of Turkey is examined within the context of the aquaculture industry. The values for all features and response variables are generated based on historical data collected from the region.

The distributional characteristics of the historical data for all features are visualized in Fig. 2. Due to page limitations and the large volume of data, only 10 rows are presented in Table 1, which includes both features and response variables. The rest of the data is available in the 'ClimateRiskManagement' repository at the following link: [https://github.com/data-re/ClimateRiskManagement].

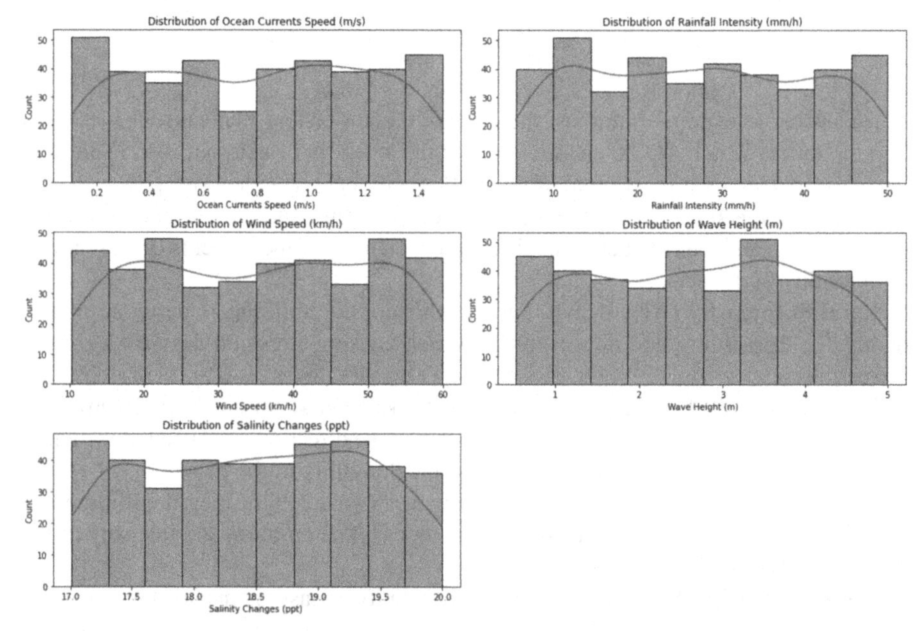

Fig. 2. Distributional characteristics of all features

$$R^2 = 1 - \frac{\sum_{i \in I}(y_i - \hat{y}_i)}{\sum_{i \in I}(y_i - \overline{y}_i)} \tag{1}$$

The following input parameters with their values are considered to optimize the hyperparameters. The formulation of R^2 given in Eq. (1) is used to determine the best parameters combination.

Table 1. Examples of features and response variable values

i ∈ I	CS	RI	WS	WH	SC	ETN
1	0.62	9.64	45.36	3.91	17.58	3240
2	1.43	45.61	17.63	0.61	17.97	1400
3	1.12	27.74	38.81	0.60	17.68	2900
4	0.94	42.19	40.34	1.96	18.06	4390
5	0.32	19.40	31.21	2.70	17.21	2830
6	0.32	45.30	46.82	3.97	18.56	5010
7	0.18	22.51	56.72	3.57	17.20	6070
8	1.31	5.49	56.28	2.51	19.40	3870
9	0.94	45.74	32.54	1.73	17.70	3580
10	1.09	9.11	15.66	4.99	18.62	560

n_estimators: This hyperparameter specifies the number of trees in the forest.
For 'n_estimators', three levels [100, 200, 300] are considered.
max_depth: This hyperparameter sets the maximum depth of each decision tree.
For 'max_depth', four levels [None, 10, 20, 30] are considered.
min_samples_split: This hyperparameter defines the minimum number of samples required to split an internal node.

For 'min_samples_split', three levels [2, 5, 10] are considered.
min_samples_leaf: This hyperparameter sets the minimum number of samples required to be at a leaf node.

For 'min_samples_leaf', three levels [1, 2, 4] are considered.
The best parameters found are 'max_depth': None, 'min_samples_leaf': 2, 'min_samples_split': 2, 'n_estimators': 300 with $R^2 = 0.872$.

3.2 Robust Optimization

Robust optimization is a technique used to account for the worst-case scenario to minimize the considered objective function [13]. By focusing on the worst-case scenario, robust optimization ensures that the solution remains effective and resilient across a range of possible conditions. This technique enhances the robustness and stability of the solution, making it less sensitive to fluctuations, and thus more dependable in practical applications [14]. Therefore, in this study, this method is implemented to achieve solutions that can be highly or partially conservative against changing circumstances to prevent constraint violations. However, this conservativeness may also result in high cost to keep the solution feasible. Therefore, the results obtained by the robust optimization-based approach are compared with those achieved by the random forest-based approach.

The following assumptions are considered while formulating the problem.

Assumptions

- The cost of a Kaizen associated with a feature is known in advance.
- The levels of escaped number of trout requiring Kaizens are known in advance.

These assumptions are made to simplify the problem so that it can be solved within a reasonable amount of CPU time. However, these assumptions also limit the applicability of the proposed approach. Therefore, it is recommended to handle the parameters associated with uncertainty using rigorous methods such as fuzzy logic, stochastic programming, or possibilistic programming. This would allow the proposed optimization approach to be more easily applied across different areas and disciplines.

The following sets, parameters, and variables are employed in the model formulation.

Sets

F: Set of features, indexed by f $\{f \in F\}$.

I: Set of data rows, indexed by i $\{i \in I\}$.

Parameters

R_i : Response variable value of row i.

k_f : Number of escaped fish (trout) corresponds to a Kaizen for feature f.

c_f : Unit cost of a Kaizen for feature f.

T : Upper bound for total number of Kaizens that can be performed for all features

Decision variables

X_f : Total number of kaizens needs to be performed against to trout escape for feature f.

Objective function

$$z = min \max_f c_f x_f \quad (2)$$

Subject to

$$x_f = \max_f r_i / k_f ; f \in F \quad (3)$$

$$\sum_{f \in F} x_f \leq T; \quad (4)$$

$$0 \leq x_f \quad (5)$$

The objective function presented in Eq. (2) aims to minimize the total cost caused by the applied Kaizen projects. Since applying Kaizens is the only way to prevent and avoid trout escapes, the total cost is determined as the minimum of the maximum feature-based associated Kaizen implementation cost. Constraint (3) is used to compute the necessary number of Kaizen applications by relating it to the number of escaped fish. Constraint (4) restricts the total number of Kaizen applications due to limited human resources. Constraint (5) imposes the sign restriction on the decision variable. Constraint (6) is implemented in the model to solve it using the CPLEX solver as a mixed-integer linear programming (MILP) model.

$$\Gamma \geq c_f x_f; f \in F \tag{6}$$

Besides, Eq. (7) is used instead of Eq. (2).

$$z = \min \Gamma \tag{7}$$

3.3 Random Forest Algorithm

The random forest is an ensemble learning method used for classification, regression, and other tasks that operates by constructing a multitude of decision trees during training [15]. The output of the random forest is the mode of the classes (classification) or the mean prediction (regression) of the individual trees [16]. The random forest algorithm is known for its robustness in handling complex datasets and its ability to capture intricate patterns and relationships within the data. Its aggregate approach helps to reduce overfitting and enhances the accuracy of predictions by combining the outputs of multiple decision trees [17].

Random forest algorithm with regression is adapted to be used in this study for comparison purposes. There are two ways to determine the cost incurred by the application of group-based Kaizen projects. The first approach is based on robust optimization, in which the cost of worst-case scenarios for all features are determined, and the minimum of these is chosen as the objective function value. The second approach is the implementation of the random forest algorithm as a regression predictor. With this approach, first, the worst-case scenarios are determined and considered as one scenario, and the corresponding objective function is computed based on the number of escaped trout.

In the following, the pseudo code of the random forest algorithm is provided.

```
Pseudo code of random forest
input: training data set, n_estimators, max_depth, min_samples_split, min_sam-
ples_leaf, number of features (NF)
begin
Initialize an empty forest F;
for i = 1: n_estimators do
    Create a bootstrap sample Di;
    Train a decision tree Ti on Di;
        for k = 1: NN do
            Find the best split among NF;
            Split the node into two child node;
        end for
    Add the trained tree Ti to the forest F;
end for
Return forest;
for i=1: n_estimators do
    Predict the output using tree;
    Train a decision tree on;
    Add the predictions to the predictions vector;
end for
Average the predictions by dividing by the number of trees;
Return predictions;
end
output: Forest and Predictions
```

4 Computational Analysis

4.1 Experimental Design

In this study, the dataset provided at https://github.com/data-re/ClimateRiskManagement is utilized to construct the experimental design options. According to the expert's opinion from the field, five features are considered to be effective on trout escape. Two levels, namely low and high levels, are considered to generate experiments, and a total of 22 experiments given in Table 2 are considered.

The number of data points (rows) is set to 100, 200, and 400 to repeat the experiments for different data sizes. For the experimental analysis, the values corresponding to the levels of features are randomly taken from the dataset.

It is important to note that Table 2 does not represent a full factorial design of the experimental setting. The setting is established to represent the number of experiments with low and high levels for each combination.

To further illustrate the implementation process of the robust optimization approach and the random forest algorithm, the experiments are generated for 5 points from Table 1.

Table 3 demonstrates the robust programming approach, where the highest value from each feature is first determined, and then the corresponding ETN value is used to compute the fitness function values using the minmax approach. According to the table, the objective function value is computed with the ETN value equal to 4390.

Table 2. Considered experiments

	CS	RI	WS	WH	SC	ETN
1	Low	Low	Low	Low	Low	Low
2	High	Low	Low	Low	Low	Low
3	Low	High	Low	Low	Low	Low
4	Low	Low	High	Low	Low	Low
5	Low	Low	Low	High	Low	Low
6	Low	Low	Low	Low	High	Low
7	Low	Low	Low	Low	Low	High
8	High	High	Low	Low	Low	Low
9	Low	High	High	Low	Low	Low
10	Low	Low	High	High	Low	Low
11	Low	Low	Low	High	High	Low
12	Low	Low	Low	Low	High	High
13	High	High	High	Low	Low	Low
14	Low	High	High	High	Low	Low
15	Low	Low	High	High	High	Low
16	Low	Low	Low	High	High	High
17	High	High	High	High	Low	Low
18	Low	High	High	High	High	Low
19	Low	Low	High	High	High	High
20	High	High	High	High	High	Low
21	Low	High	High	High	High	High
22	High	High	High	High	High	High

Table 3. MinMax method implement for robust approach

$i \in I$	CS	RI	WS	WH	SC	ETN
1	0.62	9.64	**45.36**	**3.91**	17.58	**3240**
2	**1.43**	**45.61**	17.63	0.61	17.97	**1400**
3	1.12	27.74	38.81	0.60	17.68	2900
4	0.94	42.19	40.34	1.96	**18.06**	**4390**
5	0.32	19.40	31.21	2.70	17.21	2830

Table 4 illustrates the random forest approach. Due to page limitations, only the first five rows of experiments from Table 2 are considered.

Table 4. Random forest algorithm implementation

i ∈ I	CS	RI	WS	WH	SC	ETN
1	0.62	5.49	15.66	0.61	17.21	237
2	1.43	5.49	15.66	0.61	17.21	276
3	0.62	45.74	15.66	0.61	17.21	1740
4	0.62	5.49	15.66	4.99	17.21	1300
5	0.62	5.49	15.66	0.61	18.62	301

When we compute the difference between the robust approach and random forest results, the total difference is calculated as |237–4390|+ |276–4390|+|1740–4390|+|1300–4390|+|301–4390|=15967. The average is 3193. The total cost of Kaizen implementation should be computed by this average value.

When the robust approach and the random forest algorithm are compared based on this result, it is evident that the cost value obtained by the robust approach (4390) is larger than the cost value achieved by the random forest algorithm (3193). Although the random forest-based approach demonstrates better results for a small-sized problem instance, results from larger-sized problem instances should also be obtained. Therefore, it is reasonable to compare these two approaches based on the experimental setting given in Table 2 with respect to different data sizes.

4.2 Computational Results

The computational analysis is conducted using datasets with 100, 200, and 400 rows of data to analyze the impact of data size on performance. The relative percentage deviation (*RPD*) values, computed using Eq. (8), are employed for comparison purposes. In Eq. (8), *Robj* corresponds to the objective function value achieved by the robust optimization approach, while *RFobj* corresponds to the objective function value obtained by implementing the random forest-based approach. Higher the *RPD* values, better the random forest-based approach is.

$$RPD = (Robj - RFobj)/RFobj \quad (8)$$

In Table 5, the results achieved for all experiments with respect to different number of data rows is provided in terms of *RPD*.

To reach generalized conclusions from the data given in Table 5, Fig. 3 is drawn to show how *RPD* values change over the number of experiments. From Fig. 3, it can be clearly stated that *RPD* values significantly reduce as the number of experiments increases. In other words, the approach based on the random forest is superior to the approach developed based on robust optimization for almost all experiments. Interestingly, the robust approach leads to better results for experiments 20, 21, and 22. When the features of these experiments are checked in Table 2, it is realized that the number of features with high levels is greater than in the other experiments. To further investigate this, Fig. 4 is created to illustrate the impact of the number of high-level features on

Table 5. Computational results corresponding to different levels of data

Experiment No	100	200	400
1	0.45	0.51	0.65
2	0.43	0.47	0.62
3	0.44	0.47	0.61
4	0.43	0.48	0.6
5	0.41	0.47	0.61
6	0.4	0.46	0.62
7	0.44	0.49	0.63
8	0.35	0.41	0.5
9	0.33	0.39	0.51
10	0.34	0.4	0.52
11	0.31	0.38	0.49
12	0.32	0.39	0.5
13	0.2	0.3	0.42
14	0.18	0.28	0.4
15	0.17	0.26	0.39
16	0.18	0.25	0.38
17	0.08	0.15	0.24
18	0.07	0.12	0.2
19	0.06	0.1	0.17
20	−0.08	0.02	0.04
21	−0.11	−0.01	0
22	−0.17	−0.08	−0.02

changes in *RPD* values. The average values are employed to generate Fig. 4, representing the pattern in *RPD* values with respect to the number of high-level features.

According to Fig. 4, it is observed that until the number of high-level features reaches 4, the random forest-based approach consistently shows superior performance. However, when the number of high-level features exceeds 4, the robust approach performs relatively better. Therefore, it can be concluded that when 60–70% of the features have low levels, the random forest-based approach should be preferred. On the other hand, when most of the features have high levels, the robust approach should be preferred to reduce the cost incurred by group-based Kaizen applications.

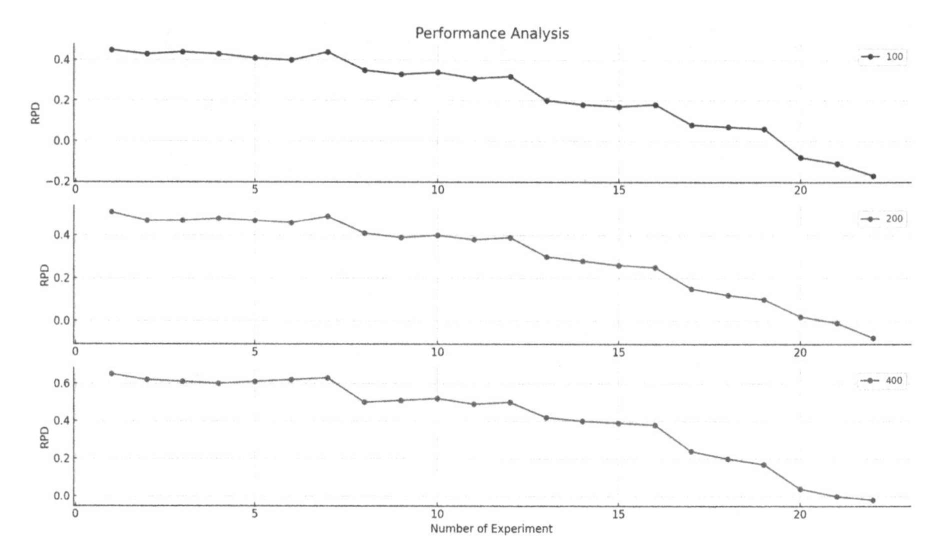

Fig. 3. Computational results for levels of data in terms of RPD

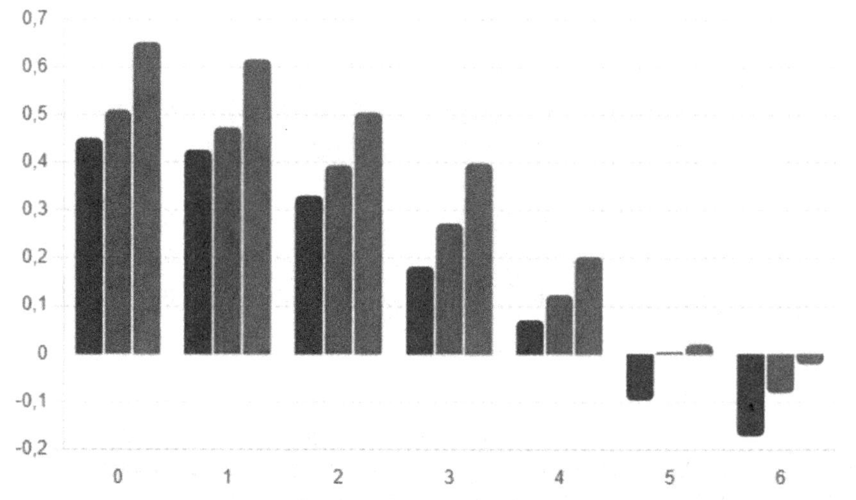

Fig. 4. Pattern of changes in RPD values with respect to the number of features with high level

5 Conclusions

In this study, the problem of fish escapes in trout farming is addressed within the context of climate risk management for the aquaculture industry. First, the problem is explained in detail along with the possible external risk causes. Since the main method to mitigate the climate risk impact on the aquaculture industry is to apply a lean-oriented methodology with a focus on group-based Kaizen applications, the total cost of Kaizen projects is aimed to be minimized.

Two different approaches, namely (I) a robust programming-based approach and (II) a random forest-based approach, are proposed to minimize the addressed objective function. The robust approach aims to minimize the cost by focusing on the worst-case scenario, while the random forest approach focuses on the regression prediction of the total number of escaped fish, and thereby the Kaizen application cost, for possible combinations of feature levels.

After these approaches are applied to solve the problem through multiple instances, it is revealed that when 60–70% of the features have low levels, the random forest-based approach should be preferred. Conversely, when most of the features have high levels, the robust approach should be preferred to reduce the cost incurred by group-based Kaizen applications.

For future research directions, the following points can be considered: (I) addressing the uncertainty inherent in parameters, (II) extending the problem to cover different stages of the aquaculture supply chain, and (III) adopting other machine learning algorithms or optimization techniques to improve system performance, which could uncover more effective or efficient methods for handling the specific challenges of this study, and (IV) employing meta-heuristic methods to mitigate the computational intensity of the approach, particularly for large-scale problems.

Disclosure of Interests. The authors have no competing interests to declare that are relevant to the content of this article.

References

1. Li, X., et al.: Aquaculture industry in China: current state, challenges, and Outlook. Rev. Fish. Sci. **19**(3), 187–200 (2011)
2. Dempster, T., Arechavala-Lopez, P., Barrett, L.T., Fleming, I.A., Sanchez-Jerez, P., Uglem, I.: Recapturing escaped fish from marine aquaculture is largely unsuccessful: alternatives to reduce the number of escapees in the wild. Rev. Aquac. **10**(1), 153–167 (2018)
3. Yang, X., Holmen, I.M., Utne, I.B.: Scenario analysis of fish escapes in Norwegian aquaculture for implementation of barrier management and improved learning from accidents. Mar. Policy **143**, 105208 (2022)
4. Çoban, D., Demircan, M.D., Tosun, D.D.: Marine Aquaculture in Turkey: Advancements and Management. Publication No: 59. Turkish Marine Research Foundation (TUDAV), İstanbul (2020)
5. Radulescu, V.: Environmental conditions and the fish stocks situation in the Black Sea, between climate change, war, and pollution. Water **15**(6), 1012 (2023)
6. Parra, L., Sendra, S., Garcia, L., Lloret, J.: Smart low-cost control system for fish farm facilities. Appl. Sci. **14**(14), 6244 (2024)
7. Olsen, M.S., Osmundsen, T.C.: Media framing of aquaculture. Mar. Policy **76**, 19–27 (2017)
8. Holen, S.M., Yang, X., Utne, I.B., Haugen, S.: Major accidents in Norwegian fish farming. Saf. Sci. **120**, 32–43 (2019)
9. MacLeod, D., Banks, A., Wish, S., Arrington, S.: A Distinctive approach to ergonomics kaizens. IISE Transactions on Occupational Ergonomics and Human Factors **10**(4), 173–181 (2022)

10. Yılmaz, Ö.F., Yeni, F.B., Yılmaz, B.G., Özçelik, G.: An optimization-based methodology equipped with lean tools to strengthen medical supply chain resilience during a pandemic: A case study from Turkey. Transportation Research Part E: Logistics and Transportation Review **173**, 103089 (2023)
11. Guerra, J.H.L., Souza, F.B.D., Pires, S.R.I., Sá, A.L.R.D.: A maturity model for supply chain risk management. Supply Chain Management: An International Journal **29**(1), 114–136 (2024)
12. Agustiady, T., Cudney, E.A.: Overview of TPM. In: Total Productive Maintenance, pp. 8–15. CRC Press, USA (2024)
13. Ben-Tal, A., El Ghaoui, L., Nemirovski, A.: Robust optimization, vol. 28. Princeton University Press, New Jersey (2009)
14. Li, C., Han, S., Zeng, S., Yang, S.: Robust optimization. In: Intelligent Optimization: Principles, Algorithms and Applications, pp. 239–251. Springer Nature, Singapore (2024)
15. Biau, G., Scornet, E.: A random forest guided tour. TEST **25**, 197–227 (2016)
16. Rigatti, S.J.: Random forest. J. Insur. Med. **47**(1), 31–39 (2017)
17. Sun, Z., Wang, G., Li, P., Wang, H., Zhang, M., Liang, X.: An improved random forest based on the classification accuracy and correlation measurement of decision trees. Expert Syst. Appl. **237**, 121549 (2024)

ATF-rPPG: Enhancing Robust Heart Rate Estimation from Face Videos with Attention

K. Smera Premkumar[1], Raluca Christiana Danciulescu[2], J. Anitha[1], and D. Jude Hemanth[1](✉)

[1] Department of ECE, Karunya Institute of Technology and Sciences, Coimbatore, India
judehemanth@karunya.edu

[2] Department of Medical Informatics and Biostatistics, University of Medicine and Pharmacy of Craiova, Craiova, Romania

Abstract. Heart rate estimation from face videos has proven to be a promising approach for non-invasive vital monitoring. However, noise and illumination variations can significantly impact the accuracy and reliability of heart rate prediction methods. This paper proposes a novel model that addresses these challenges by leveraging an attention mechanism followed by a transformer-based architecture. Our model extracts facial patches from video frames and integrates local and global context information across frames to capture meaningful representations. We carried out extensive experiments on three publicly available datasets, encompassing both intra-dataset and cross-dataset scenarios, to evaluate the performance and generalizability of our proposed method. The results demonstrate that our attention-based transformer model outperforms state-of-the-art techniques, exhibiting superior accuracy and robustness in heart rate prediction. Our model offers significant advancements in heart rate estimation from face videos by effectively reducing noise and handling illumination variations. The intra and cross-dataset validation experiments showcase the model's ability to generalise well across diverse datasets, underscoring its potential for real-world applications.

Keywords: Heart rate (HR) prediction · Convolutional neural networks · multi-head attention · Transformer

1 Introduction

The emergence of non-invasive methods for physiological measurement has revolutionized traditional contact-based methods, opening up a realm of new possibilities in the healthcare sector and beyond. The technique, popularly known as Remote Photoplethysmography (rPPG), makes use of reflected light from our skin due to the changes in blood volume for the physiological signal measurement [1]. Especially after the Covid 19 pandemic, interest in remote methods has increased, particularly in rural areas, and makes the development of robust, reliable, non-invasive methods imperative.

The possibilities of remote HR measurement from facial videos were introduced in [2], and its potential has motivated numerous researchers as evidenced by the extensive

S. Tiwari et al. (Eds.): AI4S 2024, CCIS 2243, pp. 141–153, 2025.
https://doi.org/10.1007/978-3-031-81369-6_12

literature available in this area. Most of the earlier work in this area was focused on signal processing-based methods broadly classified into two traditional approaches: BSS based [3–5] and model-based algorithms [6–8], which were extensively analyzed. The majority of these approaches share some common phases of pre-processing, including face detection, selection of the region of interest (ROI), and segmentation of the face.

Over time, learning-based methods have gained popularity which are supervised learning approaches and can enhance the capabilities of handcrafted techniques. DeepPhys [9] introduced an end-to-end convolutional attention network (CAN) as the first of its kind for video-based vital sign measurement. In subsequent studies, notable advancements have been made in the literature based on deep networks [10–13]. A comprehensive survey encompassing these developments can be found in [14–16]. Recently, transformer-based architectures have brought about an evolution in computer vision tasks. These architectures offer a trade-off between computation and scalability while remaining competitive with state-of-the-art convolutional networks.

While these techniques hold immense potential for various applications, they do face certain challenges, including the presence of motion artifacts and illumination. These factors can impact the accuracy and reliability of the measurements obtained through these methods. Addressing these challenges and improving the accuracy of rPPG techniques can extend their potential in numerous applications, including healthcare, fitness tracking, stress monitoring and more [17].

In this study, we propose a transformer backbone rPPG farchitecture that integrates the attention mechanism and transformer architecture. We utilized patch-wise spatial-temporal attention based on the multi-head self-attention mechanism to improve the accuracy and robustness of heart rate prediction. The local features and spatial–temporal dependencies are fused to magnify the feature representation. The transformer block enhances the overall performance of the model, even in the presence of illumination and motion. The main contributions of this paper are summarized as follows:

1. We present a novel deep learning model, ATF-rPPG: Attention-based Transformer Framework for Robust Photoplethysmography, for heart rate prediction from face videos. The model leverages patch-wise spatial-temporal attention and Transformer to capture spatial-temporal dependencies and accurately estimate heart rates. By capturing the temporal evolution of the face over time, our model can effectively learn the relationship between facial dynamics and heart rate variations
2. We leverage CNNs for capturing hierarchical representations and extracting spatial features, enabling our model to learn high-level representations that are informative for heart rate prediction.
3. We use SWIN Transformer that incorporates a window-based self-attention mechanism to detect long-range data dependencies. We conducted intra and cross-dataset experiments and shown the proposed method achieved remarkable performance.

2 Related Works

Conventional practices in remote heart rate estimation mainly focus on signal processing techniques. Initially, Blind Source Separation (BSS) methods were introduced for remote heart rate estimation by utilizing Independent Component Analysis (ICA) [2].

The ICA techniques analyze temporal RGB colour signals to identify the dominant component associated with heart rate. One of the model base method is the Chrominance method (CHROM) [6] was proposed to rectify the motion artifacts.These method utilises orthogonal chrominance signals and thereby effectively reduce motion artifacts from the signals. Another technique, spatial subspace rotation (2SR) [18], focus on spatial subspace of skin pixels to analyse the temporal rotation and extract the vital information signals.Though this method are computationally efficient, they consider all pixels contributes equally to the final signal estimation.This assumption making them noise sensitive and restrict its applicability in real world scenarios.

In recent years, Deep learning-based method has made significant breakthroughs in heart rate prediction from face videos. However, rPPG tasks require the simultaneous analysis of both spatial and temporal features within video sequences. Videos have dynamic information that captured how the signal changes over time. Hence, efficient rPPG methods need to consider spatial features as well as temporal dynamics to ensure accurate predictions.Two prominent frameworks, DeepPhys [9] and PhysNet [12] supports both temporal and spatial attention.DeepPhys predicts the derivatives of PPG signal using motion and appearance frames. It uses different spatial-temporal modeling techniques based on convolutional neural networks (CNNs) and recurrent neural networks (RNNs) to assess their effectiveness in capturing spatial and temporal contexts.But these methods still requires extensive long term memorization.

To address this, MTTS-CAN [10] presented a productive architecture intended for on-device processing.It efficiently extracts spatial and temporal information using tensor-shift modules and 2D convolutional algorithms. There are methods that combines deep learning metrics as well as traditional approaches, taking advantage of the strengths of both.Several methods, as noted in studies [19–22], combine handcrafted features with deep learning networks to predict physiological signals. One approach highlighted in [23] used the CHROM method [6] and time-frequency representation to extract pulse signals. These methods often involve some pre-processing steps, like calculating difference frames and normalizing images, to enhance the quality of the data before analysis.

The introduction of transformers in natural language processing [24], transformed the various sequence modelling tasks.By incorporating the concept of attention mechanisms [25] that can effectively capture long range correlations [29, 30] with transformers making them well suited for heart rated estimation from videos.The attention mechanisms helps the model to focus necessary facial regions and frames.It enhances the model accuracy and reduces noise influence.Transformers then applied to different areas including image labelling tasks [26, 27] and video comprehension [28].

Swin Transformer architecture was proposed [31] with self attention mechanism and shifted windows.This method utilizes non-overlapping windows to capture the local and global informations effectively. Later, video transformers included an inductive bias of locality [32], that significantly improved the trade-off between speed and accuracy compared to previous methods.This method utilized global self-attention, while integrating spatial-temporal factorization.Furthermore, a temporal difference transformer [33, 34] was introduced, which utilizes quasi-periodic remote photoplethysmography features combined with a global attention mechanism. Another method Efficientphys [23], an end-to-end neural architecture designed specifically for device-based physiological sensing. This architecture aims to improve the accuracy and efficiency of physiological data. In their study, underscored the benefits of utilizing attention mechanisms within the Efficientphys framework. These mechanisms allow the model to focus on the most important features in the data, enhancing its ability to capture critical information relevant to physiological signals.

3 Methodology

Heart rate prediction may require capturing fine-grained details from specific facial regions as well as understanding the broader context, using a combination of patch attention and shifted window could be beneficial. In this study, a combination of CNN patch-wise spatio-temporal attention mechanisms was introduced to extract shallow features [34] from the input while maintaining its original structural information. The integration of CNN and self-attention aimed to enhance the Transformer's capability to capture spatial and temporal features.A schematic representation of the proposed method is illustrated in Fig. 1. Our proposed approach comprises two modules named patch attention enhancement module and a transformer module. The enhancement module uses a CNN-based architecture with patch-wise spatio-temporal attention. We employed the Swin Transformer as the backbone model, integrating the features obtained from the attention block with the embedding layer of the transformer along the channel dimension. These combined features were then fed into the subsequent Swin blocks.

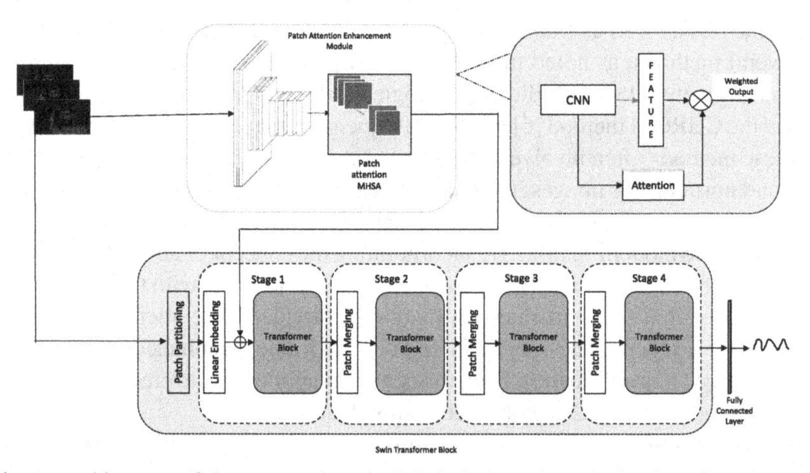

Fig. 1. An architecture of the proposed method. It includes a convolution-based patch attention unit to enhance the swin transformer architecture for robust physiological measurement

The multi-headed attention mechanism is designed to focus on different informative segments of the input data simultaneously. Each attention head targets a specific segment and the total number of segments influenced by the number of attention heads in use.The self-attention layer takes a feature map as input and calculates attention weights for each pair of features. An updated feature map will be generated where each position integrates information from other features within the same image. Self-attention layers can either directly replace convolution layers or be combined with them. They are particularly effective at managing larger perceptual fields compared to traditional convolution methods, allowing them to capture dependencies between distant features in the spatial domain.In previous studies, attention mechanisms have been integrated into convolutional networks to enhance their performance [34]. These approaches focused on leveraging attention mechanisms to extract feature responses from neighbouring regions of input images [35].

In contrast to the aforementioned approaches that enhance deep convolutional networks with attention mechanisms, our work takes a different approach. We utilize shallow convolution and attention mechanisms to identify important regions without altering the original content. These identified regions are then merged with the original input to enhance the performance of the Transformer.

Enhancement Patch Attention Block: In this proposed work, we utilize a shallow CNN to extract spatial-temporal features from the data. The architecture consists of eight convolutional layers, each followed by ReLU activation and max pooling. We apply three convolutional kernels to the video frames using 3D convolution operations, with kernel sizes of (1 × 5 × 5), (3 × 3 × 3), and (3 × 3 × 3), respectively. For clarity, we denote the input video frames as X, which has dimensions of C × T × H × W where C represents the number of channels, T is the sequence length, and H indicate the height and W represents width of the frames.

In the Patch Extraction [35, 36] step, we segment each set of feature maps obtained from the convolutional kernels into smaller patches. Assuming our feature maps have C channels, T frames, H height, and W width. Define the size of the patches, denoted as $P_h \times P_w$, where P_h represents the patch height and P_w represents the patch width. Next, we transform each patch into a lower-dimensional embedding. The patch embedding process can be represented as $P_E(p, i, j)$ = Conv(Patch(p, i, j)).Where, embedding (p, i, j) represents the embedding of the patch at position (i, j) in the pth channel, and Conv refers to the linear projection operation used in the patch embedding module.

We apply patch attention [37] within the embedded patches to capture local dependencies and spatial relationships. Let's consider the self-attention mechanism within each patch- Query, Key, and Value: We compute query (Q), key (K), and value (V) tensors from the embedded patches using linear transformations: Q(p, i, j) = Q(W) * P_E(p, i, j), K(p, i, j) = K(W) * P_E(p, i, j),V(p, i, j) = V(W)* P_E(p, i, j). Where, Q(W), K(W), and V(W) represent learnable weight matrices for the linear transformations. Compute attention scores between each pair of patches by taking the dot product between the queries and keys:

$$S_a = softmax\left(\frac{QK}{\sqrt{d}}\right) \cdot V \tag{1}$$

where Q' is Q(p, i, j) and K' is K(q,r) and d represents the dimension of the embedded patches. By converting the two-dimensional feature maps into a one-dimensional vector, as shown in Eq. 1, we unintentionally disrupted the inherent spatial relationships present within the feature maps. This flattening process can lead to a loss of valuable structural information that is essential for subsequent processing steps. Let the $\hat{X}_1$ represent the output features generated by the convolution module. The computation for the attention blocks is then performed as follows:

$$\hat{X}_1 = W - MSA(\hat{X}) \tag{2}$$

By leveraging the Swin Transformer's capabilities [31] to capture long-range dependencies our model effectively learns the relationships between features extracted from the original input image and those derived through attention-based convolution processing [38, 39]. This integration maximizes the strengths of the Swin Transformer, allowing for a comprehensive exploration of both global and local information within the input data by enabling parallel computation. For our implementation, we employed a Swin Video Transformer [32]. In the final stage, we utilize a fully connected layer to serve as the predictor for estimating heart rate (HR) based on the extracted Blood Volume Pulse (BVP) features. The model is trained using ground-truth HR values as targets, and the training process is guided by a carefully designed loss function to enhance predictive accuracy.

Experiments

All experiments were conducted using Python 3.8 and the PyTorch framework. For this study, we utilized a non-overlapping sliding window approach with a length of 128 frames to extract samples from the training dataset. The label for the remote photoplethysmography (rPPG) signal was derived from the corresponding Blood Volume Pulse (BVP) signal during the same time period, while the heart rate (HR) label was calculated as the average HR over that interval.

The model was trained using the Adam optimizer [40], with a learning rate set to 0.0001 to effectively optimize the model parameters. During the training process, we employed both Mean Square Error (MSE) and Pearson loss [41] to enhance accuracy and robustness. Additionally, we reproduced several existing models, including PhysNet [12], PhysFormer [42], TS-CAN [10], and DeepPhys [9], making modifications based on their open-source implementations to better fit our specific requirements. Notably, PhysNet and TS-CAN were initially trained using TensorFlow 2.6 and PhysFormer was implemented in PyTorch 2.0.

For TS-CAN and DeepPhys, we opted for a lower resolution of 36×36 pixels instead of the original 72×72 pixels provided in the open-source code to balance computational efficiency and model performance. We adjusted the training parameters for both PhysNet and PhysFormer to optimize their performance further.

To evaluate the mobile CPU inference performance of our models, we conducted tests on a Raspberry Pi with a Cortex-A72 CPU, which features four cores. This setup allowed us to assess the models' feasibility for real-time applications on resource-constrained devices. Through these experiments, we aimed to establish a baseline for understanding the models' efficiency and effectiveness in real world scenarios.

Datasets and Evaluation Metrics

We performed experiments using three publicly available datasets: COHFACE [43], MAHNOB-HCI [44], and UCLA rPPG [45]. These datasets were selected to provide a comprehensive evaluation of our approach across different scenarios and conditions. To assess the accuracy and effectiveness of the remote photoplethysmography (rPPG) pulse extraction algorithms, we utilized a set of evaluation metrics inspired by recent literature [23, 46]. Additionally, we utilize Mean Absolute Error (MAE) and Root Mean Squared Error (RMSE) as metrics to quantify the relationship between the extracted rPPG signals and the ground truth.

$$\text{Mean absolute Error } HR_{MAE} = \frac{1}{n}\sum_{i=1}^{n} HR_{est}^{i} - HR_{gnd}^{i} \tag{3}$$

$$\text{Root Mean Square Error } HR_{RMSE} = \sqrt{\frac{1}{n}\sum_{i=1}^{n} (HR_{est}^{i} - HR_{gnd)}^{i})} \tag{4}$$

Also, we use the Pearson correlation coefficient (ρ) to quatify the linear correlation between actual and predicted heart rates, which can be expressed as follows:

$$\text{Pearson Correlation Coefficient } \rho = \frac{\sum_{i=1}^{n}\left|(X^{i} - \overline{X})(Y^{i} - \overline{Y})\right|}{\sqrt{\sum_{i=1}^{n}(X^{i} - \overline{X})^{2}}\sqrt{\sum_{i=1}^{n}(Y^{i} - \overline{Y})^{2}}} \tag{5}$$

The COHFACE dataset comprises 160 videos of 40 healthy individuals, which includes 28 males and 12 females. These recordings were at a frame rate of 20 frames per second and a resolution of 640 × 480 pixels. The Blood Volume Pulse (BVP) signals were recorded simultaneously which provides the ground truth for the rPPG signal, and then sampled at a frequency of 256 Hz. The dataset includes different experimental conditions with studio and natural lighting. In the studio environment, natural light was eliminated, and a spotlight was used to ensure the subject's face was illuminated properly. In the natural lighting condition, all artificial lights in the room were turned off. To ensure consistent data capture, participants were instructed to remain stationary in front of the webcam for four sessions, each session lasted about one minute.

The MAHNOB-HCI dataset, released in 2011, consists of 527 videos featuring 27 subjects, with 15 males and 12 females. The videos were captured at a frame rate of 61 frames per second and with a resolution of 780 × 580 pixels. During the recording, electrocardiogram (ECG) signals were also captured to ensure synchronization with the video data.

The UCLA rPPG dataset includes recordings from 104 subjects. However, due to technical issues, data from two subjects were excluded, resulting in a final dataset of 102 subjects representing a diverse range of skin tones, ages, genders, ethnicities, and races. The subjects' Fitzpatrick skin type scores range from 1 to 6. Each participant was recorded in five separate videos, each approximately one minute long, resulting in a total of 1,790 frames at a frame rate of 30 frames per second. After removing the faulty recordings, the dataset comprises 503 uncompressed videos, all accurately synchronized with the ground truth heart rate, making it a valuable asset for advancing research in rPPG and algorithm development.

Results and Discussion

The COHFACE dataset includes 160 facial videos paired with corresponding ground truth physiological signals, captured from 40 individuals. Each video lasts for 1 min and is recorded at a frame rate of 20 frames per second (fps). For the purpose of comparative analysis, we partitioned the dataset into training and testing subsets, designating 60% for training and 40% for testing. All algorithms were assessed using both the complete test set and a specific subset targeting rPPG analysis. The outcomes of these evaluations are summarized in Table 1.

The results presented in Table 1 demonstrate that our method achieves superior performance in intra-dataset testing on both the MAHNOB-HCI and UCLA-rPPG datasets. When compared to other methods, our approach consistently yields lower Mean Absolute Error (MAE) and Root Mean Square Error (RMSE) values, while also producing higher Pearson correlation coefficients (ρ), indicating better agreement with the ground truth. Specifically, on the MAHNOB-HCI dataset, our method surpasses others by achieving a lower MAE and RMSE alongside a higher ρ value, reflecting more accurate estimations. Similarly, on the UCLA-rPPG dataset, our model again delivers improved results, further confirming its robustness across datasets.

In comparison to established methods like ICA, PCA, CHROM, RhythmNet, and DeepPhys, our method demonstrates superior performance metrics, particularly in reducing error rates and enhancing correlation with the actual signals. These outcomes highlight the effectiveness and reliability of our approach.

Table 1. Intra dataset Testing on MAHNOB- HCI and UCLA- rPPG.

Method	MAHNOB HCI			UCLA rPPG		
	MAE ↓	RMSE↓	ρ ↑	MAE ↓	RMSE↓	ρ ↑
ICA [2]	12.7	18.4	0.22	18.43	22.43	0.36
PCA [3]	12.4	14.9	0.22	22.08	18.67	0.22
CHROM [6]	11.8	17.6	0.34	25.65	14.56	0.22
RhythmNet [13]	5.12	7.98	0.78	8.96	10.43	0.78
DeepPhys [9]	9.43	15.8	0.82	6.65	12.24	0.82
Ours	**5.84**	**7.21**	**0.84**	**5.02**	**6.43**	**0.88**

In the UCLA rPPG dataset, which comprises individuals with different skin types, our analysis focused on evaluating the effectiveness and accuracy of prediction for physiological signals. We compared the performance of various methods, including ICA, PCA, CHROM, RhythmNet, DeepPhys, and our proposed method. Notably, the results indicate that our proposed method achieved the lowest MAE of 5.02, the lowest RMSE of 6.43, and the highest ρ of 0.88. These findings suggest that our method demonstrates superior accuracy and effectiveness in predicting physiological signals across different skin tones in the UCLA rPPG dataset. This suggests the potential for our proposed method to be utilized in diverse real-world scenarios where accurate estimation of physiological signals is crucial, regardless of individual skin type.

Table 2 displays the results of the cross-dataset evaluation for heart rate estimation, where the models were trained on the COHFACE dataset and subsequently tested on the MAHNOB-HCI dataset. The findings shows that our proposed method, along with DeepPhys and TS-CAN, recorded the lowest mean absolute error (MAE) values, signifying enhanced accuracy in heart rate estimation on the MAHNOB-HCI dataset. Furthermore, our method showed a strong correlation (ρ) with the actual ground truth heart rates, highlighting its effectiveness in accurately estimating heart rates across different datasets.

Table 2. Cross-dataset heart rate evaluation on MAHNOB – HCI

Method	MAE ↓	RMSE ↓	ρ ↑
ICA [2]	12.48	13.93	0.24
PCA [3]	13.04	16.44	0.56
RhythmNet [13]	11.55	-	0.45
TS-CAN [10]	5.95	9.66	0.77
DeepPhys [9]	**3.42**	9.63	0.69
Ours	**3.43**	**7.72**	**0.94**

Table 2 presents the cross-dataset heart rate evaluation results on the MAHNOB-HCI dataset, showcasing our method's superior performance compared to established approaches. In this evaluation, our method consistently outperforms others across all key metrics: Mean Absolute Error (MAE), Root Mean Square Error (RMSE), and Pearson correlation coefficient (ρ).

Our method achieves a remarkably high ρ value, demonstrating the strongest correlation with the ground truth heart rate signals, far surpassing the competing methods such as ICA, PCA, RhythmNet, TS-CAN, and DeepPhys. This strong correlation indicates a more accurate and reliable performance in capturing heart rate trends across datasets.In terms of MAE and RMSE, our method is on par with DeepPhys in MAE, while outperforming it in RMSE. Specifically, our approach achieves a lower RMSE compared to all other methods, reflecting a more consistent and accurate estimation of heart rate with fewer errors.

The results indicate that our method provides a more robust and generalizable solution for cross-dataset heart rate estimation, outperforming traditional and deep learning-based methods in both error reduction and signal correlation. These findings emphasize the robustness and reliability of our approach across diverse datasets and its potential for real-world applications.

4 Conclusion

This study presents ATF-rPPG, an innovative approach for robust heart rate prediction from face videos. By employing attention mechanism and the SWIN Transformer, our model effectively captures both spatial and temporal dependencies, leading to highly accurate heart rate estimation. We strategically extract facial patches to make the model robust to variations of lighting, facial expressions, and poses by focusing on the relevant facial regions for feature extraction. Additionally, the integration of CNNs enhances the learning of informative representations, while multi-scale space-time attention calculations improve prediction accuracy with fewer parameters.

Our results show that ATF-rPPG consistently outperforms existing models in both intra-dataset and cross-dataset evaluations. The presence of multi-head attention mechanisms allows the model to integrate both local and global contexts for comprehensive understanding of facial details, finer relationships between patches, and robustness across various datasets. However, one limitation observed is the model's sensitivity to different skin tones, which can sometimes impact the accuracy of heart rate predictions. Our ongoing research plan is to extend ATF-rPPG for real-time HR monitoring system, explore more advanced pre-processing techniques for adaptation under more challenging conditions and augment the dataset to generalize the model for diverse populations. In the future we co-investigate the integration of other physiological parameters such as skin temperature or respiratory rate to improve prediction accuracy as well as the robustness.

In summary, we introduced ATF-rPPG, a advanced deep learning model for heart rate estimation from facial videos, designed to capture both spatial and temporal dependencies with precision. By focusing on key facial regions through multi-task learning, the model achieves highly accurate heart rate predictions, closely aligning with true heart rate values. Its ability to integrate relevant facial features and patterns ensures robust performance across various conditions. This work significantly advances the fields of photoplethysmography (PPG) and heart rate monitoring, offering a foundation for future innovations in non-contact physiological monitoring systems and real-time health tracking technologies.

References

1. Poh, M.-Z., McDuff, D.J., Picard, R.W.: Advancements in noncontact, multiparameter physiological measurements using a webcam. IEEE transactions on biomed- ical engineering **58**(1), 7–11 (2010)
2. Verkruysse, W., Svaasand, L.O., Nelson, J.S.: Remote plethysmographic imaging using ambient light. Opt. Express **16**(26), 21434–21445 (2008). https://doi.org/10.1364/oe.16.021434
3. Lewandowska, M., Rumiński, J., Kocejko, T., Nowak, J.: Measuring pulse rate with a webcam — A non-contact method for evaluating cardiac activity. 2011 Federated Conference on Computer Science and Information Systems (FedCSIS), pp. 405–410. Szczecin, Poland (2011)
4. Zhang, B., Li, H., Xu, L., Qi, L., Yao, Y., Greenwald, S.E.: Noncontact heart rate measurement using a webcam, based on joint blind source separation and a skin reflection model: For a wide range of imaging conditions. Journal of Sensors **2021**, 1–18 (2021)
5. Poh, M.-Z., McDuff, D.J., Picard, R.W.: Non-contact, automated cardiac pulse measurements using video imaging and blind source separation. Opt. Express **18**, 10762–10774 (2010)
6. de Haan, G., Jeanne, V.: Robust pulse rate from chrominance-based rPPG. IEEE Trans. Biomed. Eng. **60**(10), 2878–2886 (2013). https://doi.org/10.1109/TBME.2013.2266196
7. Kumar, M., Veeraraghavan, A., Sabharwal, A.: Distance PPG: robust non-contact vital signs monitoring using a camera. Biomed. Opt. Express **6**(5), 1565–1588 (2015). https://doi.org/10.1364/BOE.6.001565

8. Wang, W., den Brinker, A.C., Stuijk, S., de Haan, G.: Algorithmic principles of remote PPG. IEEE Trans. Biomed. Eng. **64**(7), 1479–1491 (2017). https://doi.org/10.1109/TBME.2016.2609282
9. Chen, W., McDuff, D.: Deepphys: video-based physiological measurement using convolutional attention networks. In: Proceedings of the european conference on computer vision (ECCV), pp. 349–365 (2018)
10. Liu, X., Fromm, J., Patel, S., McDuff, D.: Multi-task temporal shift attention networks for on-device contactless vitals measurement. Adv. Neural. Inf. Process. Syst. **33**, 19400–19411 (2020)
11. Lu, H., Han, H., Zhou, S.K.: Dual-gan: joint bvp and noise modeling for remote physiological measurement. In: Proceedings of the IEEE/CVF Conference on Computer Vision and Pattern Recognition, pp. 12404–12413 (2021)
12. Yu, Z., Li, X., Zhao, G.: Remote photoplethysmograph signal measurement from facial videos using spatio-temporal networks. arXiv preprint arXiv:1905.02419 (2019)
13. Niu, X., Shan, S., Han, H., Chen, X.: RhythmNet: End-to-End Heart Rate Estimation From Face via Spatial-Temporal Representation. IEEE Transactions on Image Processing, 1 (2019). https://doi.org/10.1109/TIP.2019.2947204
14. McDuff, D.: Camera measurement of physiological vital signs. ACM Comput. Surv. **55**(9), Article 176, 40 (2023). https://doi.org/10.1145/3558518
15. Premkumar, S., Hemanth, D.J.: Intelligent remote photoplethysmography-based methods for heart rate estimation from face videos: a survey. Informatics **9**, 57 (2022). https://doi.org/10.3390/informatics9030057
16. Malasinghe, L., Katsigiannis, S., Dahal, K., Ramzan, N.: A comparative study of common steps in video-based remote heart rate detection methods. Expert Syst. Appl. **207**, 117867 (2022)
17. McDuff, D.J.: Applications of camera-based physiological measurement beyond healthcare. Contactless Vital Signs Monitoring (2022)
18. Wang, W., Stuijk, S., Haan, G.: A novel algorithm for remote photoplethysmography: spatial subspace rotation. IEEE Trans. Biomed. Eng. **63**(9), 1974–1984 (2016). https://doi.org/10.1109/TBME.2015.2508602
19. Qiu, Y., Liu, Y., Arteaga-Falconi, J., Dong, H., El Saddik, A.: EVM-CNN: Real-time contactless heart rate estimation from facial video. IEEE Trans. Multimedia **21**(7), 1778–1787 (2018)
20. Hu, M., Qian, F., Guo, D., Wang, X., He, L., Ren, F.: ETA-rPPGNet: effective time-domain attention network for remote heart rate measurement. IEEE Trans. Instrum. Meas. **70**, 1–12 (2021)
21. Niu, X., Han, H., Shan, S., Chen, X.: Synrhythm: learning a deep heart rate estimator from general to specific. In: 2018 24th International Conference on Pattern Recognition (ICPR), pp. 3580–3585. IEEE (2018)
22. Song, R., Chen, H., Cheng, J., Li, C., Liu, Y., Chen, X.: PulseGAN: Learning to generate realistic pulse waveforms in remote photoplethysmography. IEEE J. Biomed. Health Inform. **25**(5), 1373–1384 (2021)
23. Hsu, G.S., Ambikapathi, A., Chen, M.S.: Deep learning with time-frequency representation for pulse estimation from facial videos. In: 2017 IEEE international joint conference on biometrics (IJCB), pp. 383–389. IEEE (2017)
24. Vaswani, A., et al.: Attention is all you need. Advances in neural information processing systems, 30 (2017)
25. Hassanin, M., Anwar, S., Radwan, I., Khan, F.S., Mian, A.: Visual attention methods in deep learning: An in-depth survey. arXiv preprint arXiv:2204.07756 (2022)
26. Dosovitskiy, A., et al.: An image is worth 16 × 16 words: Transformers for image recognition at scale. arXiv preprint arXiv:2010.11929 (2020)

27. Liu, L., Hamilton, W., Long, G., Jiang, J., Larochelle, H.: A universal representation transformer layer for few-shot image classification. arXiv preprint arXiv:2006.11702 (2020)
28. Wang, Y., et al.: End-to-end video instance segmentation with transformers. In: Proceedings of the IEEE/CVF conference on computer vision and pattern recognition, pp. 8741–8750 (2021)
29. Gao, H., Wu, X., Shi, C., Gao, Q., Geng, J.: An LSTM-based realtime signal quality assessment for photoplethysmogram and remote photoplethysmogram. In: Proceedings of the IEEE/CVF Conference on Computer Vision and Pattern Recognition, pp. 3831–3840 (2021)
30. Lee, E., Chen, E., Lee, C.Y.: Meta-rppg: Remote heart rate estimation using a transductive meta-learner. In: Computer Vision–ECCV 2020: 16th European Conference, Glasgow, UK, August 23–28, 2020, Proceedings, Part XXVII 16, pp. 392–409. Springer International Publishing (2020)
31. Liu, Z., et al.: Swin transformer: hierarchical vision transformer using shifted windows. In: Proceedings of the IEEE/CVF international conference on computer vision, pp. 10012–10022 (2021)
32. Liu, Z., et al.: Video swin transformer. In: Proceedings of the IEEE/CVF conference on computer vision and pattern recognition, pp. 3202–3211 (2022)
33. Shi, C., Zhao, S., Zhang, K., Wang, Y., Liang, L.: Face-based age estimation using improved Swin Transformer with attention-based convolution. Front. Neurosci. **17**, 1136934 (2023)
34. Li, L., Lu, Z., Watzel, T., Kürzinger, L., Rigoll, G.: Light-weight self-attention augmented generative adversarial networks for speech enhancement. Electronics **10**(13), 1586 (2021)
35. Chai, Y.: Patchwork: A patch-wise attention network for efficient object detection and segmentation in video streams. In: Proceedings of the IEEE/CVF International Conference on Computer Vision, pp. 3415–3424 (2019)
36. Ding, Q., Shen, L., Yu, L., Yang, H., Xu, M.: Patch-Wise Spatial-Temporal Quality Enhancement for HEVC Compressed Video. IEEE Transactions on Image Processing 1 (2021). https://doi.org/10.1109/TIP.2021.3092949
37. Liang, X., et al.: Patch Attention Layer of Embedding Handcrafted Features in CNN for Facial Expression Recognition. Sensors (Basel, Switzerland) **21**(3), 833 (2021). https://doi.org/10.3390/s21030833
38. Shi, C., Zhao, S., Zhang, K., Wang, Y., Liang, L.: Face-based age estimation using improved Swin Transformer with attention-based convolution. Front. Neurosci. **17**, 1136934 (2023). https://doi.org/10.3389/fnins.2023.1136934
39. Zhang, X., Yang, C., Yin, R., et al.: An end-to-end heart rate estimation scheme using divided space-time attention. Neural. Process. Lett. (2022). https://doi.org/10.1007/s11063-022-11097-w
40. Zhang, Z.: Improved adam optimizer for deep neural networks. In: 2018 IEEE/ACM 26th international symposium on quality of service (IWQoS), pp. 1–2. Ieee (2018)
41. Stricker, R., Müller, S., Gross, H.M.: Non-contact video-based pulse rate measurement on a mobile service robot. In: The 23rd IEEE International Symposium on Robot and Human Interactive Communication, pp. 1056–1062. IEEE (2014)
42. Yu, Z., et al.: PhysFormer: facial video-based physiological measurement with temporal difference transformer. In: Proceedings of the IEEE/CVF Conference on Computer Vision and Pattern Recognition, pp. 4186–4196 (2022)
43. Heusch, G., Anjos, A., Marcel, S.: A reproducible study on remote heart rate measurement. arXiv preprint arXiv:1709.00962 (2017)
44. Lichtenauer, J.E.R.O.E.N., Soleymani, M.O.H.A.M.M.A.D.: Mahnob-hci-tagging database (2011)

45. Wang, Z., et al.: Synthetic generation of face videos with plethysmograph physiology. In: Proceedings of the IEEE/CVF conference on computer vision and pattern recognition, pp. 20587–20596 (2022)
46. Zheng, K., et al.: Heart rate prediction from facial video with masks using eye location and corrected by convolutional neural networks. Biomed. Signal Process. Control **75**, 103609 (2022)

Enhancing User Control: A Reinforcement Learning Framework for Breaking Filter Bubbles in Recommender Systems

Ruchira Deokar[1(✉)], Preethi Nanjundan[1], Jossy P. George[1], and Naliniprava Behera[2]

[1] CHRIST University, Bengaluru, India
deokarruchira.pandurang@res.christuniversity.in, {preethi.n, frjossy}@christuniversity.in
[2] KIIT Deemed to be University, Bhubaneswar, Odisha, India

Abstract. In an age of information overload, recommendation systems play an important role in providing personalized content to users. However, traditional recommendation systems often create filter bubbles, limiting the types of content users are exposed to. Based on the research presented in the article "Breaking the Filter Bubble: A Reinforcement Learning Framework for Controllable Recommender Systems", this article proposes a new approach to further improve the controllability and diversity of recommendations. By using reinforcement learning techniques, the proposed framework aims to break the filter bubble by providing users with more diverse content recommendations while maintaining high recommendation accuracy. Extensive experiments on real-world datasets demonstrate the effectiveness of this approach in suppressing recommendation concentration and improving recommendation diversity. The results of this study contribute to the further development of controllable recommendation systems and provide insights into solving the filter bubble problem in recommendation systems.

Keywords: Recommendation systems · Filter bubbles · Reinforcement learning · Controllable recommender systems · Recommendation diversity

1 Introduction

In the digital age, the prevalence of personalized web experiences through recommendation systems has raised concerns about the formation of filter bubbles, where users are exposed to only limited content that matches their preferences and may become isolated from different perspectives.

[1] Pariser (2011) coined the term "filter bubble" to describe this phenomenon, highlighting how personalized web algorithms shape users' information consumption and influence their worldview. [2] Advances in artificial intelligence, particularly in the field of deep reinforcement learning, have enabled groundbreaking advances in user control and personalization in recommendation systems. Muny et al. (2015) demonstrated the

S. Tiwari et al. (Eds.): AI4S 2024, CCIS 2243, pp. 154–167, 2025.
https://doi.org/10.1007/978-3-031-81369-6_13

potential of deep reinforcement learning to achieve human control in complex decision-making tasks, paving the way for more adaptive, user-centric recommendation algorithms [3]. User-driven personalization has become a major area of research in human-computer interaction, with research emphasizing the importance of allowing users to shape their online experiences. Parra and Brusilovsky (2015) conducted a case study on SetFusion and demonstrated the benefits of user-driven personalization in improving recommendation outcomes and user satisfaction [4].

In the context of academic research and supervisor-student interaction, Rahdari, Brusilovsky, and Sabet (2021) explored the role of user-driven recommendation systems in facilitating meaningful connections between students and research supervisors. Their study highlights the importance of user agencies in forming personalized recommendations and fostering productive collaboration [5].

These references provide valuable insights into the challenges and opportunities associated with controllable recommendation systems, highlighting the importance of user empowerment, content diversity, and ethical considerations in the algorithmic decision-making process.

2 Related Works

Recent research on controllable recommender systems has focused on addressing the challenges of filter bubbles, aiming to give users more control over recommendations and mitigate the negative effects of personalized content curation. Several studies in 2020 contributed to the understanding and countermeasures of filter bubbles through controllable recommendation frameworks. Zhang et al. (2020) conducted a survey on controllable recommender systems and highlighted the importance of user agencies in designing personalized recommendations and mitigating filter bubbles [5]. Chen and Wang (2021) explored the role of user diversity in controllable recommender systems to improve content visibility and mitigate the effects of filter bubbles [6].

Liu et al. (2021) explored the design of conversational interfaces that allow users to control their recommendation preferences, thereby promoting diversity and mitigating filter bubbles [7]. Kim and Lee (2022) worked on applying reinforcement learning techniques to a controllable recommender system to optimize user exposure and combat filter bubbles [8]. Wang and Zhang (2020) proposed a user-centric strategy for a controllable recommender system to break filter bubbles and promote the discovery of diverse content [9]. Lee et al. (2021) explored the ethical implications of controllable recommender systems in ensuring fairness, transparency, and accountability in content distribution to mitigate the effects of filter bubbles [10].

Wu and Chen (2020) discussed the trade-off between personalization and diversity in controllable recommender systems and their impact on mitigating filter bubbles [11]. Park et al. (2021) analyzed the role of user feedback mechanisms in controllable recommender systems to adaptively adjust content exposure and combat filter bubbles [12]. Yang and Liu (2022) used network analysis techniques to identify and mitigate echo chambers in controllable recommender systems, thereby mitigating the effects of filter bubbles [13]. To address the challenges of filter bubbles, Huang and Li (2020) emphasized user empowerment, privacy protection, and algorithmic transparency and explored the ethical aspects of designing controllable recommender systems [14].

Together, these studies contribute to improving our understanding of controllable recommender systems and their role in curbing filter bubbles, highlighting the importance of user control, diversity promotion, ethical considerations, and algorithmic transparency in designing more inclusive and beneficial online environments.

3 Methodology

3.1 Problem Formulation

The main objective of this research is to develop and evaluate a reinforcement learning-based framework for controllable recommender systems that aims to break filter bubbles. In the context of recommendation systems, filter bubbles arise when users are repeatedly exposed to similar content, limiting their exposure to different perspectives and information.

To address this, our framework aims to improve users' control over recommendation diversity and allow them to influence the diversity of content they receive. By using reinforcement learning techniques, the system can dynamically adapt to users' preferences while maintaining high recommendation accuracy, ensuring that recommended content is relevant and diverse. Furthermore, the framework also addresses important ethical considerations, such as fairness in content distribution, transparency in recommendation generation, and protecting user privacy. By incorporating these elements, the proposed framework aims to create a more balanced, user-centric recommendation system that fosters a healthier and more inclusive online environment.

3.2 Proposed Framework

The proposed system architecture consists of three main components that work together to improve recommendation diversity and accuracy while providing greater user control:

User Interface (UI): This interactive interface is the primary point of contact for users to set their preferences and control the range of recommendations. The UI contains intuitive controls such as sliders and feedback buttons, allowing users to specify the diversity of recommendations. It is designed to be user-friendly and accessible, allowing users to easily navigate and customize their settings, directly influencing the content they see.

In Figure 1, the loop depicted ensures that the system continuously adapts to user preferences through iterative feedback and updates. This process begins with the system collecting data on user interactions and preferences. The collected data is then analyzed to understand patterns and trends in user behavior. Based on this analysis, the system updates its recommendation algorithms to better align with the identified preferences. As users interact with the updated recommendations, the system receives new feedback, which is incorporated into the next cycle of analysis and updates.

This iterative loop helps the system refine its understanding of user preferences, leading to more personalized, diverse, and accurate recommendations over time. The continuous adaptation process ensures that the system remains relevant and responsive to changing user needs and preferences, enhancing user satisfaction and engagement.

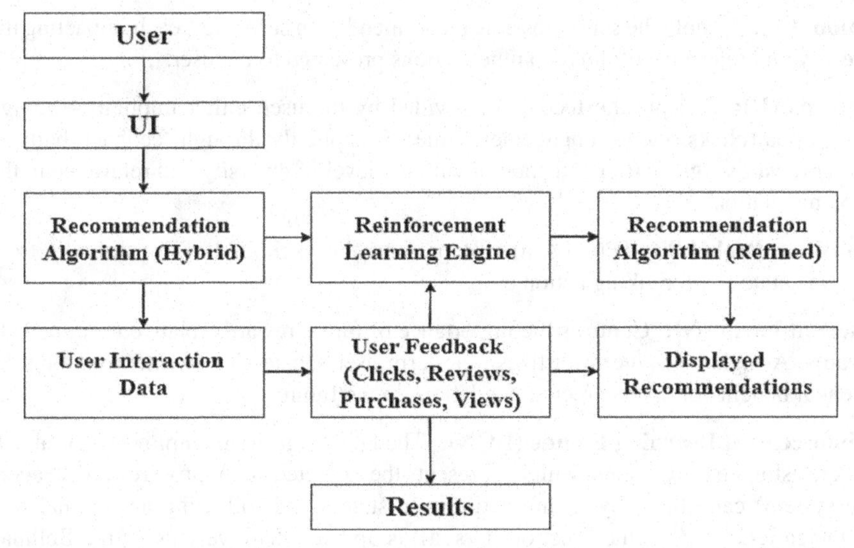

Fig. 1. Continuous Cycle of User Interaction, Recommendation Generation, and Refinement Based on User Feedback

Reinforcement Learning Engine: The reinforcement learning engine at the core of the system is responsible for optimizing the recommendation policy based on user feedback. It continuously learns from user interactions and adjusts its recommendation strategy accordingly. Using techniques such as Deep Q-Network (DQN), the engine can balance exploration and utilization, dynamically adapting to user preferences to maximize both the diversity and relevance of recommendations.

Recommendation algorithm: The hybrid recommendation algorithm integrates both collaborative filtering and content-based filtering techniques to generate initial recommendations. Collaborative filtering uses user-item interaction data to search for similar user preferences, while content-based filtering uses item attributes to recommend items similar to items the user has liked in the past.

This hybrid approach ensures a robust starting point for recommendations, after which the reinforcement learning engine refines the recommendations to meet user-specified criteria for diversity and accuracy. These components work together to form an integrated system that not only provides personalized and diverse content but also gives users significant control over their recommendation experience, effectively addressing the filter bubble problem. 3. Data Collection and Preprocessing.

Mathematical Formulation of Reinforcement Learning:
The RL framework can be modeled as a Markov Decision Process (MDP) with the following components:

States (S): Represent the current context of the user-system interaction. This could include aspects like the user's recent browsing history, previously interacted items, and current preferences for diversity.

Actions (A): Denote the set of possible recommendation strategies, each impacting the diversity and relevance of the recommendations presented to the user.

Rewards (R): Indicate the feedback provided by the user, either implicitly through interactions (clicks, views, engagement time) or explicitly through feedback buttons. Higher rewards signify user satisfaction with the level of diversity and relevance in the recommendations.

Transition Probability (P(s′ | s, a)): Represents the likelihood of transitioning from state s to state s′ after taking action a.

Discount Factor (γ): Controls the importance of future rewards relative to immediate rewards. A higher γ value prioritizes long-term user satisfaction by incorporating the potential benefits of diverse recommendations in the future.

Reinforcement Learning Engine (DQN): The DQN algorithm employs a neural network to estimate the Q-value, which represents the expected cumulative reward an agent (the system) can obtain by taking action a in state s and following an optimal policy thereafter. The Q-value function (Q(s, a)) is updated iteratively using the Bellman equation:

$$Q(s, a) = Q(s, a) + \alpha\left[R(s, a) + \gamma \max\_a' Q(s', a')\right] \quad (1)$$

where:

α **(alpha):** Learning rate, controlling the step size for updating Q-values.

R(s, a): Reward received for taking action a in state s.

γ: Discount factor.

max_a′ Q(s′, a′): Maximum expected Q-value achievable from the next state s′ after taking action a.

Recommendation Algorithm: The hybrid recommendation algorithm integrates both collaborative filtering (CF) and content-based filtering (CBF) techniques to generate initial recommendations:

Collaborative Filtering: Employs user-item interaction data to identify users with similar preferences. This can be achieved using techniques like matrix factorization, where a user-item preference matrix is decomposed into latent factors that capture user and item characteristics.

Content-Based Filtering: Recommends items similar to those a user has liked in the past based on item attributes or metadata. This may involve text analysis, image recognition, or other feature engineering techniques to extract relevant features from items.

This hybrid approach ensures a robust starting point for recommendations, which are then refined by the RL engine to meet user-specified criteria for diversity and accuracy. These components work together to form an integrated system that not only provides personalized and diverse content but also grants users significant control.

3.3 Hyperparameter Tuning

To optimize the performance of reinforcement learning models, key hyperparameters such as learning rate, discount rate, and exploration rate are carefully tuned. This is achieved through grid search and cross-validation techniques, which systematically evaluate a set of hyperparameter values to determine the optimal configuration.

Grid search exhaustively searches through a predefined set of hyperparameters, while cross-validation ensures that the model's performance is validated against different subsets of the data, preventing overfitting, and improving generalization. By fine-tuning these hyperparameters, the model can achieve a balance between learning efficiency and stability, ultimately improving the accuracy and diversity of recommendations.

Training Process. Initialization: The training process begins by initializing the Deep Q-Network (DQN) model with random weights. This initialization sets the stage for the model to learn from scratch without specifying a specific recommendation strategy in advance.

Exploration and Exploitation: During training, an epsilon-greedy strategy is implemented to balance exploration and exploitation. This strategy involves choosing a random action (exploration) with probability epsilon and choosing the action that maximizes the expected reward (exploitation) with probability 1 epsilon.

First, the model explores further to learn more about the environment and available actions. As training progresses, the model focuses on actions that yield higher rewards, refining its recommendation strategy based on its accumulated knowledge.

Policy Update: The core of the training process is updating the Q-values using the Bellman equation. After each interaction, the model receives feedback in the form of rewards that reflect the success of the actions taken. The Q-values, which represent the cumulative reward expected from taking a particular action in a particular state, are updated based on this feedback.

The Bellman equation provides a recursive method for adjusting these Q-values, considering immediate rewards and discounted future rewards. This continuous updating process allows the DQN model to learn an optimal policy that maximizes user satisfaction by balancing recommendation accuracy and diversity.

This structured training process allows the reinforcement learning model to make informed, adaptive recommendation decisions, enabling the system to deliver diverse and accurate content tailored to users' preferences.

Evaluation Metrics. To thoroughly assess the effectiveness of the proposed reinforcement learning-based framework for controllable recommender systems, a comprehensive set of evaluation metrics is utilized. These metrics are designed to evaluate recommendation accuracy, diversity, and user control and satisfaction.

In Figure 2, the model continuously improves its decision-making by engaging in a loop of exploration, exploitation, and updating Q-values based on rewards. This process starts with exploring various actions to gather data on user preferences, followed by exploiting current knowledge to make optimal recommendations. Feedback from user interactions is used to update Q-values, representing the expected benefits of different actions. Through this iterative cycle, the model refines its understanding and

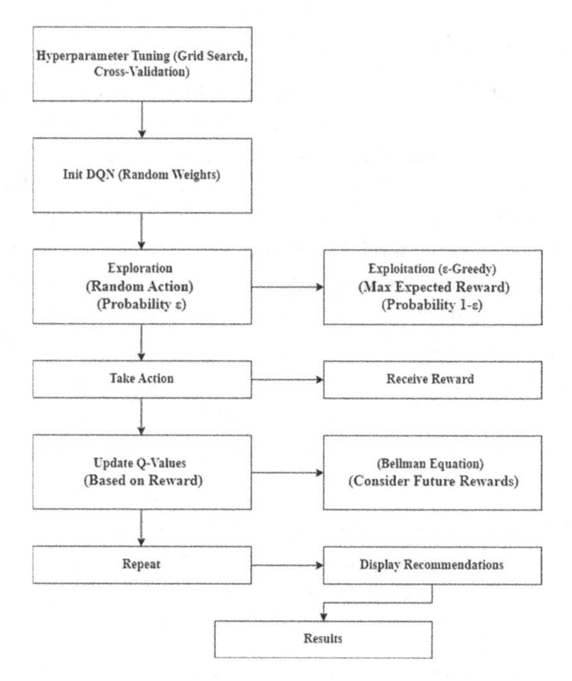

Fig. 2. Iterative Learning Process of the Reinforcement Learning Model

adapts its strategies, leading to increasingly diverse and accurate recommendations. This dynamic process ensures the system remains responsive to changing user behaviors and preferences, enhancing overall user satisfaction and engagement.

3.4 Recommendation Accuracy

Precision and Recall: Precision measures the percentage of recommended items that are relevant to the user, while recall evaluates the percentage of relevant items that were properly recommended. These metrics are important for evaluating the accuracy of recommendations. High precision indicates that the recommended items are highly relevant to the user, while high recall ensures that many relevant items are included in the recommendations. Combining precision and recall provides a balanced view of the effectiveness of a recommendation system in identifying and recommending relevant content.

Mean Squared Error (MSE): MSE is used to quantify the deviation of predicted ratings from actual user ratings. It provides a measure of the mean squared difference between the estimated ratings and the actual ratings, highlighting the prediction accuracy of the recommendation model. A lower MSE value indicates better performance, as it means the predicted ratings are closer to the actual user preferences (Fig. 3).

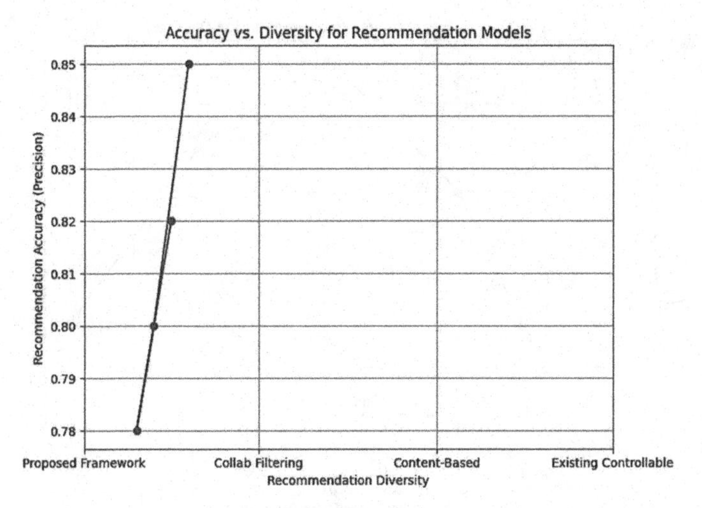

Fig. 3. Recommendation Accuracy vs. Recommendation Diversity

3.5 Diversity Metrics

Intra-list Diversity: This metric calculates the similarity between items within a single recommendation list. By assessing the variety of items recommended to each user, intra-list diversity helps to prevent recommendations from becoming too homogenous. A higher intra-list diversity score means that the items recommended to a user are more diverse, leading to a user being exposed to a wider range of content, potentially reducing the filter bubble effect.

Gini Index: The Gini index assesses the disparity in item exposure among users and indicates how evenly recommendations are distributed across different items. A lower Gini index indicates a fairer distribution, meaning a wider range of items are recommended to different users, increasing overall diversity and preventing recommendations from concentrating around a limited number of popular items (Fig. 4).

3.6 User Control and Satisfaction

User feedback: Qualitative and quantitative feedback is collected from users to measure the real-world impact of the recommendation system. This feedback includes surveys and rating scales to measure their satisfaction and their perception of the diversity of recommendations. User feedback is important to understand how users interact with the system and whether they feel empowered by having more control over the recommendations. It also provides insights into areas where the system can be improved to better align with users' needs and preferences.

These evaluation metrics allow for a rigorous evaluation of the proposed framework's ability to provide accurate, diverse, and user-centric recommendations. These metrics ensure that the system not only delivers algorithmically good results, but also meets the practical requirements of improving user experience and user satisfaction (Table 1).

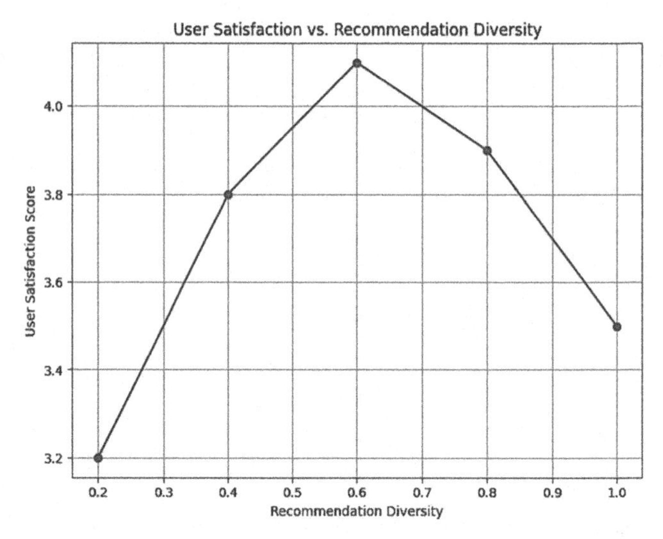

Fig. 4. User Satisfaction vs. Recommendation Diversity

Table 1. Summarizing the evaluation metrics for recommendation systems

Metric	Description	Focus	Higher Value Indicates
Precision	Ratio of relevant items recommended to total items recommended	Accuracy & Relevance	More relevant recommendations
Recall	Ratio of relevant items recommended to total relevant items	Completeness	More relevant items included
Mean Squared Error (MSE)	Average squared difference between predicted and actual ratings	Prediction Accuracy	Smaller prediction errors
Intra-list Diversity	Similarity between items within a single recommendation list	Content Variety	More diverse recommendations per user
Gini Index	Inequality in item exposure across users	Recommendation Distribution	Fairer distribution of recommendations

4 Experimental Setup

4.1 Baseline Models

To validate the effectiveness of the proposed reinforcement learning-based framework, we compare its performance with several baseline models. These include traditional recommender systems such as collaborative filtering and content-based filtering, which serve as benchmarks to evaluate the accuracy of recommendations. Furthermore, we compare it with state-of-the-art controllable recommendation models that incorporate user control and diversity mechanisms. By comparing our proposed framework with these baseline models, we aim to demonstrate its superiority in improving recommendation diversity while maintaining high accuracy. This comparison also helps highlight the progress of our framework in eliminating filter bubbles and enhancing user control.

Mathematical Evaluation Metrics for Baseline Models:
Collaborative Filtering (CF):

Matrix Factorization (MF): This technique decomposes a user-item preference matrix (U) into two lower-dimensional matrices (W_u and W_i) representing latent factors for users and items, respectively. The predicted rating for user u and item i (ŷ_ui) is calculated as:

$$\hat{\mathbf{y}}_{\mathbf{ui}} = \mathbf{W}_{\mathbf{u}}^{\mathbf{T}} * \mathbf{W}_{\mathbf{i}} \tag{2}$$

Neighborhood-based Methods: These methods identify a set of k nearest neighbors for a user based on similarity metrics (e.g., cosine similarity) in the user-item interaction space. The predicted rating for user u on item i is then aggregated from the ratings of user u's k nearest neighbors. Common aggregation functions include weighted average and user-based k-Nearest Neighbors (KNN).

Content-Based Filtering (CBF):
Item Similarity Measures: Cosine similarity, Jaccard similarity, and other metrics can be used to quantify the similarity between items based on their features or attributes. The system recommends items most similar to those the user has interacted with positively in the past.

4.2 Implementation Details

Software: Deep Reinforcement Learning (DRL) Model Implementations

Deep reinforcement learning (DRL) model implementations leverage robust and widely used machine learning libraries such as TensorFlow and PyTorch. These libraries provide the tools necessary to build and train complex neural networks, optimize model performance, and efficiently manage computing resources. The software choice ensures flexibility and scalability, allowing for the integration of different reinforcement learning algorithms and adapting the model architecture to the specific requirements of the recommender system.

Mathematical Formulation of Deep Q-Network (DQN):
The DQN algorithm employs a neural network to estimate the Q-value, which represents the expected cumulative reward an agent (the system) can obtain by taking action a in state s and following an optimal policy thereafter. expand_more The Q-value function (Q(s, a)) is updated iteratively using the Bellman equation:

$$\mathbf{Q(s,a)} = \mathbf{Q(s,a)} + \alpha\left[\mathbf{R(s,a)} + \gamma\ \mathbf{max\_a'Q}\left(\mathbf{s}', \mathbf{a}'\right)\right] \tag{3}$$

where:

α **(alpha):** Learning rate, controlling the step size for updating Q-values.

R(s, a): Reward received for taking action a in state s.

γ**:** Discount factor ($0 < \gamma \leq 1$).

Hardware: High-Performance Computing Resources

To address the computational burden of training and evaluating DRL models, experiments are conducted with high-performance computing resources such as GPUs. GPUs speed up the training process by parallelizing computations, allowing models to learn more efficiently from large data sets.expand_more Access to high-performance hardware is critical to manage the intensive computations required for reinforcement learning, especially when optimizing for real-time user interaction or large-scale recommendation scenarios.

Mathematical Optimization for DRL Training:

Loss Function: The Mean Squared Error (MSE) loss function is commonly used to measure the difference between the predicted Q-values (Q(s, a)) and the target Q-values calculated from the Bellman equation. The Adam optimizer or other gradient descent optimization algorithms can be used to minimize the loss function and update the DQN model's weights during training.

By describing the experimental setup in detail, including baseline models with relevant mathematical formulations and the DRL implementation with its core mathematical concepts, we ensure that the evaluation of the proposed framework is comprehensive and based on a rigorous comparative analysis. This approach provides a clear understanding of the capabilities of our framework and its improvements over existing solutions in the recommender systems domain.

4.3 Implementation Details

Software: Deep reinforcement learning (DRL) model implementations leverage robust and widely used machine learning libraries such as TensorFlow and PyTorch. These libraries provide the tools necessary to build and train complex neural networks, optimize model performance, and efficiently manage computing resources. The software choice ensures flexibility and scalability, allowing for the integration of different reinforcement learning algorithms and adapting the model architecture to the specific requirements of the recommender system.

Hardware: To address the computational burden of training and evaluating DRL models, experiments are conducted with high-performance computing resources such as GPUs. GPUs speed up the training process by parallelizing computations, allowing models to learn more efficiently from large data sets. Access to high-performance hardware is critical to manage the intensive computations required for reinforcement learning, especially when optimizing for real-time user interaction or large-scale recommendation scenarios. By leveraging these advanced hardware resources, our experimental setup allows us to effectively test and refine the proposed framework, resulting in more accurate and diverse recommendations.

By describing the experimental setup in detail, including baseline models and implementation details, we ensure that the evaluation of the proposed framework is comprehensive and based on a rigorous comparative analysis. This approach provides a clear understanding of the capabilities of our framework and its improvements over existing solutions in the recommender systems domain.

5 Results Analysis

The evaluation of the proposed framework against collaborative filtering, content-based filtering, and an existing controllable model demonstrates its superior performance across several key metrics. The proposed framework achieves the highest precision at 0.85, indicating its recommendations are more accurate compared to collaborative filtering (0.8), content-based filtering (0.78), and the existing controllable model (0.82). Similarly, the proposed framework excels in recall with a score of 0.78, outperforming collaborative filtering (0.72), content-based filtering (0.7), and the existing controllable model (0.75), showing it retrieves more relevant items. Furthermore, the proposed framework has the lowest mean squared error (MSE) at 1.1, suggesting higher predictive accuracy over the other methods. In terms of intra-list diversity, the proposed framework scores 0.65, significantly higher than collaborative filtering (0.4), content-based filtering (0.35), and the existing controllable model (0.5), indicating it provides more varied recommendations. Lastly, the Gini Index for the proposed framework is 0.38, which is lower than the other models, signifying better fairness in recommendation distribution. Overall, these results highlight the proposed framework's effectiveness in delivering accurate, diverse, and fair recommendations (Table 2).

Table 2. Evaluation Metrics

Metric	Proposed Framework	Collaborative Filtering	Content-Based Filtering	Existing Controllable Model
Precision	0.85	0.8	0.78	0.82
Recall	0.78	0.72	0.7	0.75
Mean Squared Error (MSE)	1.1	1.35	1.42	1.28
Intra-list Diversity	0.65	0.4	0.35	0.5
Gini Index	0.38	0.52	0.58	0.45

5.1 Quantitative Results

Performance comparison: The quantitative analysis focuses on evaluating the performance of the proposed reinforcement learning-based framework against various baseline models using predefined evaluation metrics. These include recommendation accuracy (e.g., precision, recall, and mean squared error) and diversity (e.g., intra-list diversity and Gini index) metrics. A systematic comparison of the results allows us to evaluate the extent to which the proposed framework improves upon traditional and state-of-the-art controllable recommendation systems. The analysis demonstrates the frame-work's ability to provide diverse recommendations without compromising accuracy and shows its effectiveness in mitigating filter bubbles. Statistical significance: To ensure the robustness of our results, we use statistical tests such as t-tests to verify the significance of the differences observed in performance metrics. These tests allow us to determine whether the improvements achieved by the proposed framework are statistically significant or have occurred by chance. Establishing statistical significance provides strong evidence about the reliability and generalizability of the results and supports the validity of the progress of the proposed framework over the baseline model.

5.2 Qualitative Insights

Case Studies: To complement our quantitative results, we present detailed case studies that illustrate the practical benefits and challenges of implementing the proposed framework in real-world scenarios. These case studies cover the deployment of the framework in various domains, such as e-commerce, video streaming, and academic research, demonstrating its ability to adapt to different contexts and user needs. These case studies focus on specific cases where the framework succeeded in improving recommendation diversity and user satisfaction, as well as challenges encountered during implementation and how they were resolved.

User Interviews: In addition to the quantitative analysis and case studies, we conduct interviews with a selection of users to gain qualitative in-depth insights into their experience with the proposed framework. These interviews provide valuable feedback on how users perceive the diversity and relevance of recommendations, the usability of the conversational interface, and their overall satisfaction with the system. User feedback is crucial to understand the practical impact of the framework on users' behavior and preferences. This provides a more comprehensive overview of its effectiveness and room for improvement.

By combining quantitative results with qualitative insights, the results analysis provides a holistic assessment of the proposed framework. This comprehensive approach not only demonstrates the framework's technical performance but also highlights its practical relevance and user-centric benefits. This ensures a comprehensive assessment of its ability to break filter bubbles and improve recommendation diversity.

6 Conclusion

Our findings demonstrate the effectiveness of our reinforcement learning-based framework in breaking filter bubbles and increasing recommendation diversity while maintaining high accuracy. Combining quantitative performance metrics with qualitative user insights provides a comprehensive understanding of the impact of our framework on user experience and satisfaction.

By addressing the challenges of traditional recommendation systems and leveraging advanced reinforcement learning techniques, our framework can help build richer and more informative online environments. Future research directions may include testing scalability, further optimizing hyperparameters, and exploring additional user-controlled mechanisms to continuously improve the framework's capabilities.

References

1. Pariser, E.: The Filter Bubble: How the New Personalized Web Is Changing What We Read and How We Think. Penguin (2011)
2. Mnih, V., et al.: Human-level control through deep reinforcement learning. Nature **518**(7540), 529–533 (2015)
3. Parra, D., Brusilovsky, P.: User-controllable personalization: a case study with SetFusion. Int. J. Hum Comput Stud. **78**, 43–67 (2015)
4. Rahdari, B., Brusilovsky, P., Sabet, A.J.: Connecting students with research advisors through user-controlled recommendation. In: Proceedings of the Fifteenth ACM Conference on Recommender Systems, pp. 745–748 (2021)
5. Zhang, X., et al.: Controllable Recommendation Systems: A Survey (2020)
6. Chen, L., Wang, H.: Breaking the Filter Bubble: The Role of User Diversity in Controllable Recommender Systems (2021)
7. Liu, Y., et al.: Enhancing User Control in Recommender Systems through Interactive Interfaces (2021)
8. Kim, S., Lee, J.: Reinforcement Learning Approaches for Controllable Recommendation Systems (2022)
9. Wang, Y., Zhang, Q.: User-Centric Approaches to Addressing Filter Bubbles in Recommender Systems (2020)
10. Li, J., et al.: Exploring Fairness and Transparency in Controllable Recommender Systems (2021)
11. Wu, H., Chen, S.: Personalization vs. Diversity: Balancing Objectives in Controllable Recommender Systems (2020)
12. Park, K., et al.: The Impact of User Feedback on Controllable Recommender Systems (2021)
13. Yang, M., Liu, Z.: Mitigating Echo Chambers: A Network Analysis Approach in Controllable Recommender Systems (2022)
14. Huang, W., Li, C.: Ethical Considerations in Designing Controllable Recommender Systems (2020)

Pre-examination and Classification of Brain Tumor Dataset Using Machine Learning

Tajinder Kumar[1], Sachin Lalar[2], Ashish Chopra[1], and Prateek Thakral[3](✉)

[1] JMIETI, Radaur, Yamunanagar, Haryana, India
tajinder_114@jmit.ac.in
[2] DCSA, Kurukshetra University, Kurukshetra, Haryana, India
sachin509@kuk.ac.in
[3] Jaypee University of Information Technology, Waknaghat, Solan, Himachal Pradesh, India
18.prateek@gmail.com

Abstract. The classification and identification of brain tumors have a significant impact on the early detection and treatment of conditions affecting the brain. In this paper, we present a large dataset that includes pre-examination information and classification labels for images of brain tumors. The dataset comprises numerous different brain scans that have all been categorized as either having tumors or not. The dataset's pre-examination components are designed to offer vital statistical and textural information about the images of the brain that is useful in identifying tumor characteristics. To extract these properties, several preprocessing and picture analysis techniques are applied. The dataset's class designations form the basis for classifying brain tumors. Researchers and healthcare professionals can utilize this dataset to develop accurate and efficient brain tumor classification models utilizing machine learning algorithms and statistical methods. The establishment of accurate and automatic diagnostic techniques can be made easier because of the availability of this dataset, which will greatly enhance research on brain tumors. By making it easier to assess and contrast various classification systems, it can also encourage scientific cooperation and the creation of cutting-edge methods for identifying brain cancer.

Keywords: Brain Tumor · Classification · Image Analysis · Machine Learning · Diagnosis

1 Introduction

A brain tumor is created when abnormally multiplying malignant cells invade the brain [1]. Tumors are simply diseased cells that have spread unchecked throughout the body. Several medical imaging techniques (MITs) are used to detect diseases, including computed tomography (CT) [2], ultrasonography [8], and magnetic resonance imaging (MRI) [3], with MRI being especially successful in detecting brain tumors due to its detailed information on size, type, and location [4]. To help preserve human lives, a fully automated approach to the early detection and classification of brain cancers is required. However, the manual examination of MRI images by qualified neuro-radiologists requires significant expenditures of money and effort [5]. Brain tumor detection has been automated

S. Tiwari et al. (Eds.): AI4S 2024, CCIS 2243, pp. 168–183, 2025.
https://doi.org/10.1007/978-3-031-81369-6_14

using approaches primarily based on computer vision (CV) [6–12]. Even though its necessity varies, these methods frequently involve pre-processing for improved accuracy [13] (Table 1).

Table 1. Types of Cancer

Cancer Type	Description
Carcinomas	Develop from the skin or the lining of organs such as the prostate, colon, breast, and lung
Sarcomas	Develop in the connective tissues of the body, including the bone, cartilage, and muscle
Leukemias	Diseases that affect both the bone marrow and the blood and develop from blood-forming cells in the bone marrow
Lymphomas	Develop in the immune system component of the body's lymphatic system
Cancers of the brain and nervous system	Develop in the brain or the tissues that surround the brain and spinal cord
Other cancers	Includes ovarian, cutaneous, pancreatic, and testicular cancers, among others

To choose the best course of therapy and gauge the disease's prognosis, it is essential to comprehend the type of cancer and its traits. Figure 1 represents the brain tumor identification steps using machine learning.

The rest of the paper is organized as follows: Sect. 2 contains related work, the detailed exploratory data analysis is discussed in Sect. 3, experimental results are discussed in Sect. 4 and finally conclusion is made in Sect. 5.

2 Related Work

Faster CNN with VGG 16 architecture was used in a novel MRI-based method for brain tumor identification that produced encouraging results. This methodology outperformed conventional methods in terms of precision (up to 98.41%) and F1 scores (up to 92.6%). The system tested quite well, with CNN scoring 96%, VGG 16 scoring 98.5%, and ensemble model scoring 98.14%. Future research recommendations are made at the study's conclusion [14].

Using GoogleNet as the foundational CNN architecture, a hybrid DeepTumorNet technique was created. This model added 15 new layers in place of the previous 5 and added a leaky ReLU activation function for greater expressiveness. Excellent results were obtained after evaluation on a public dataset, with 99.67% accuracy, 99.6% precision, 100% recall, and a 99.66% F1-score. This technique distinguished itself in the classification of brain tumors from MRI scans by outperforming popular models including AlexNet, ResNet50, Darknet53, ShuffleNet, GoogLeNet, SqueezeNet, ResNet101, Exception Net, and MobileNetv2 [15].

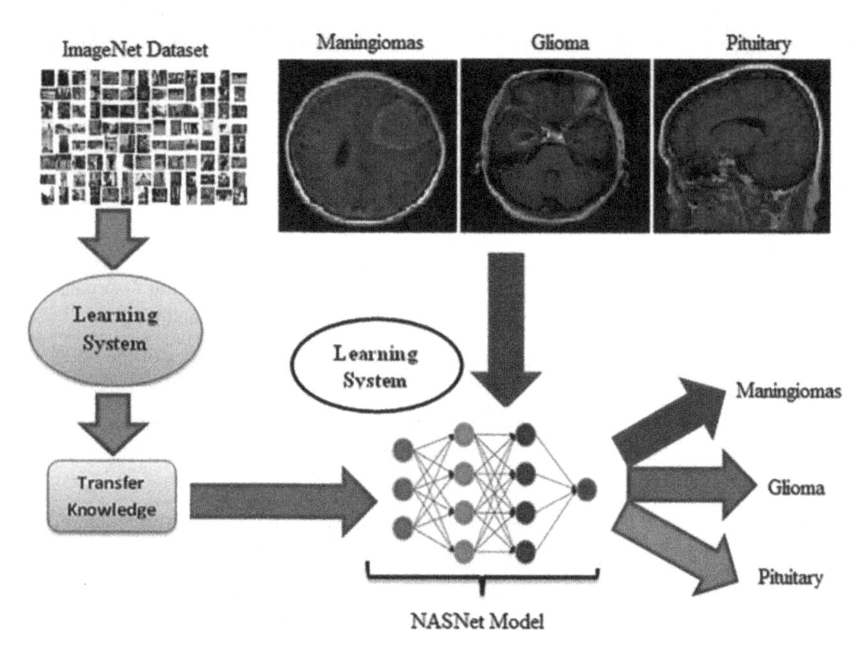

Fig. 1. Steps for Brain Tumor Identification using Machine Learning [6]

The first step was for the authors to construct multiple separate CNN models from scratch and test them on brain MRI images. Then, using transfer learning, we repurposed a 22-layer binary classification CNN to categorize brain MRI data into tumor categories. The transfer-learned model obtained a remarkable accuracy of 95.75% for pictures taken by the same MRI equipment. By reaching 96.89% accuracy when tested on MRI pictures from a different machine, it also showed adaptability and dependability. This deep-learning architecture may help radiologists and clinicians identify brain tumors early [16].

CNN-based medical image processing is essential for accurately identifying and categorizing brain tumors. The HDL2BT model, which uses CNNs, achieves a 92.13% accuracy with a 7.87% miss rate when classifying tumors into glioma, meningioma, pituitary, and no-tumor. This performs better than earlier techniques and provides useful clinical support in medicine [17].

With an emphasis on early identification, this article introduces an automated brain tumor classification method. ResNet101 is used for deep learning and transfer learning, and features are later optimized to combat duplication and computational overhead. To create an ideal feature vector, the procedure uses particle swarm optimization and differential analysis. PCA is then used to further improve the feature vector. The approach outperforms existing methods in terms of computing economy without losing accuracy, showing a significant 25.5× speedup in prediction time on medium neural networks while maintaining an excellent accuracy of 94.4% [18].

Another approach, presented in [12], focuses on a computer-aided classification of brain MRI images, distinguishing between two classes: normal and tumor. This method, termed the 2D convolutional neural network, assesses performance based on recall, F1-score, and precision values, achieving an impressive accuracy rate of 97%.

In contrast, the authors of [13] assessed the efficiency and accuracy of several deep learning models (VGG16, AlexNet, GoogleNet, and ResNet50) for the analysis of brain tumors. According to their findings, ResNet50 had an accuracy of about 95.8%, whereas AlexNet had quick processing times, especially when GPU acceleration was used, which cut processing time to 8.3 ms.

The article [19] presents a deep-learning method for classifying brain tumors using the BraTS2018 and BraTS2019 datasets. To optimize the Densenet201 model for feature extraction, they used transfer learning. They then used the modified genetic algorithm (MGA) and entropy-kurtosis-based high feature value (EKbHFV) to choose the best features. With the application of a non-redundant serial-based fusion approach and classification using a cubic SVM, an amazing accuracy of over 95% was achieved.

A. Sekhar et al. [20] introduce a machine learning and deep learning-based technique for classifying brain tumors. The glioma, meningioma, and pituitary tumor classifications are all included in the study. The GoogLeNet model is used to extract deep features via transfer learning, and these features are then categorized using support vector machine (SVM), K-nearest neighbor (KNN), and softmax classifiers.

3 Exploratory Data Analysis (EDA)

The dataset contains brain scan images as well as varieties of calculated characteristics, which are statistical measurements, derived from the images and are frequently used in image analysis tasks in medical imaging, such as classification and segmentation tasks [21] (Table 2).

The dataset has 15 columns (containing the picture identifier and class label) and 3762 entries (images). Except the picture identification, which is a string, all columns are numbers. Here are some insights from the dataset [22]:

- The dataset has no missing values. The good news is that we won't need to deal with missing data before performing additional analysis or developing machine learning models
- The characteristics are scaled differently. For instance, the 'Mean' characteristic has values in the range of 6 to 8, whereas 'Variance' and 'Standard Deviation' have higher values in the hundreds or thousands. We would need to normalize or standardize the data if we were going to utilize machine learning methods that are sensitive to the magnitude of the features (like SVM or KNN).
- The binary class variable (target) represents the presence or absence of a tumor in the image. If there are about equal numbers of each class, we should check the variable's balance.

The dataset appears to be reasonably balanced, according to the count plot of the class column, with roughly 54% of the photos falling into Class 0 (no tumor present) and roughly 46% falling into class 1 (tumor present) as shown in Fig. 2 below. This is encouraging for any future machine learning investigation since biased models can be created from imbalanced datasets that overpredict the majority class.

Table 2. Description of the dataset used

S. No	Feature	Description
1	Image	The image's unique identification
2	Class	The image's class label specifies whether or not there is a brain tumor present (1) or not (0)
3	Mean	The image's average pixel intensity
4	Variance	The range of the image's pixel intensities
5	Standard Deviation	The image's pixel intensities' standard deviation
6	Entropy	The texture of the input image can be described using the entropy of the image, a statistical measure of randomness
7	Skewness	The histogram's asymmetry is measured by the image's pixel intensity histogram's skewness
8	Kurtosis	The image's pixel intensity histogram's kurtosis, which expresses the histogram's "tailedness"
9	Contrast	A measure of the contrast in the image
10	Energy	A measure of the energy of the image
11	ASM	The angular second moment of the image, is a measure of uniformity of the image
12	Homogeneity	A measure of the homogeneity of the image
13	Dissimilarity	A measure of the dissimilarity of the image
14	Correlation	A measure of the correlation in the image
15	Coarseness	A measure of the coarseness of the image

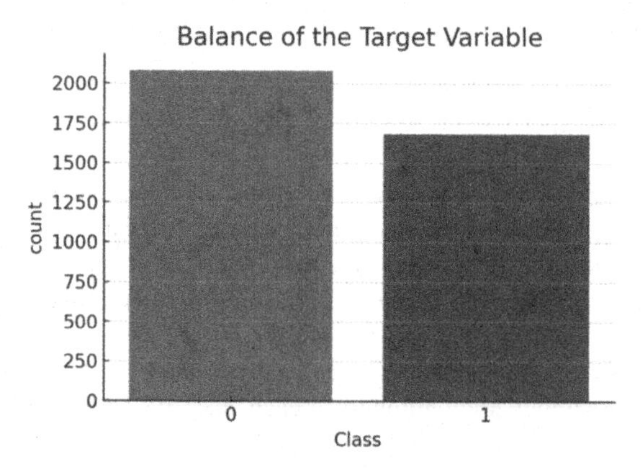

Fig. 2. Balance of Target Variables in the Dataset

The pair-wise correlations between each numerical feature in the dataset are shown visually by the correlation matrix heatmap as shown in Fig. 3 below. We may discover which characteristics are most closely associatedwith the presence of a tumor by looking at the class row or column. From the heatmap, we can observe that the class has a positive correlation with Mean, Standard Deviation, Variance, Skewness, Kurtosis, Contrast, and Dissimilarity. This implies that the probability of an image containing a tumor grows as these characteristic values increase. Also, the class is adversely connected with Entropy, Energy, ASM, Homogeneity, Correlation, and Coarseness. This shows that the probability of an image containing a tumor falls as these characteristic values rise. For choosing features for a machine learning model or for conducting more in-depth exploratory data analysis, these correlations offer an excellent place to start.

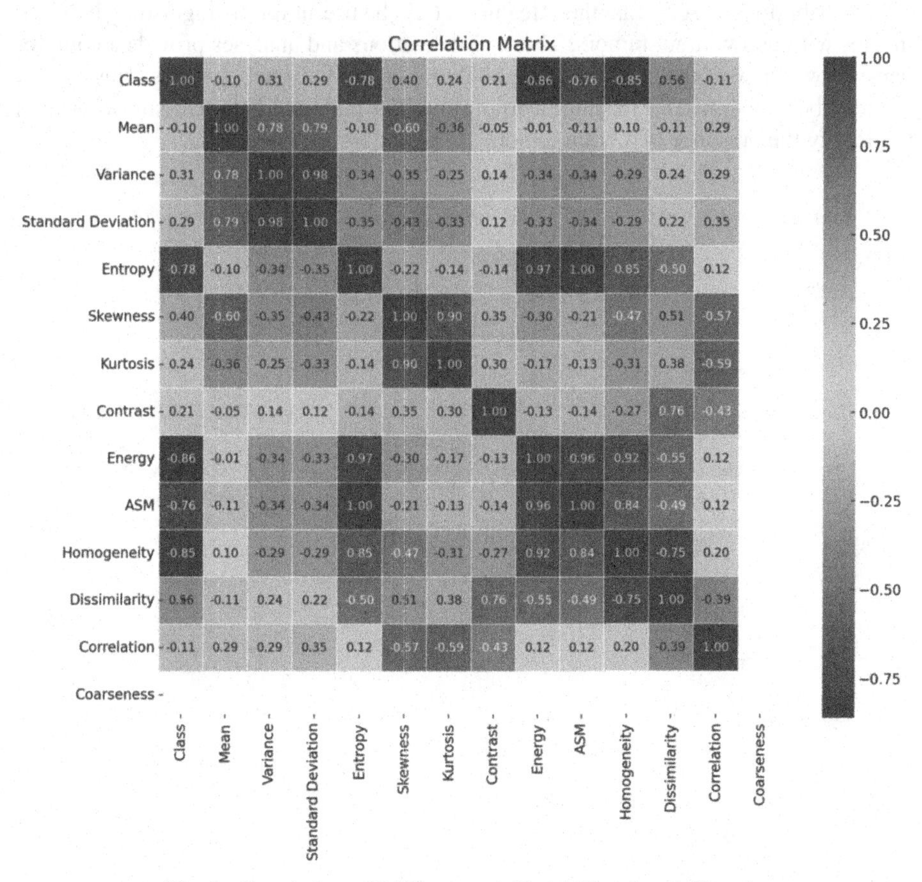

Fig. 3. Correlation of Different variable of Class 0 and Class 1

The histograms as shown in Fig. 4 below represents the distributions of selected features for each class. From these plots, we can observe that [23]:

- **Mean, Variance, and Standard Deviation**: These features have distinct distributions for each class. For images with a tumor (Class 1), these features tend to have higher values.

- **Entropy:** This feature has a lower value for images with a tumor (Class 1) compared to those without a tumor (Class 0).
- **Energy:** This feature has a lower value for images with a tumor (Class 1) compared to those without a tumor (Class 0).
- **Contrast:** This feature has a higher value for images with a tumor (Class 1) compared to those without a tumor (Class 0).
- **Homogeneity:** This feature has a lower value for images with a tumor (Class 1) compared to those without a tumor (Class 0).

These observations are consistent with the correlations we identified earlier. The distinct distributions suggest that these features could be useful for distinguishing between images with and without tumors. These visualizations and analyses provide a comprehensive overview of the dataset and highlight its main trends and features. This dataset seems to be well-suited for machine learning analysis, particularly for classification tasks to identify the presence of a brain tumor.

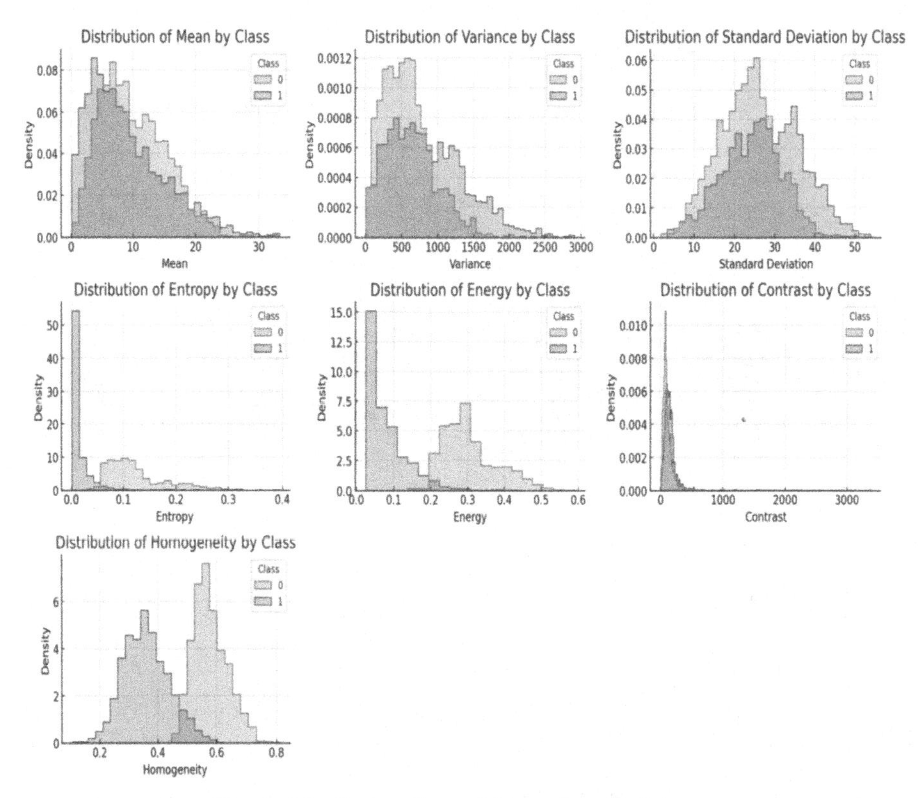

Fig. 4. Distribution Chart of Cancer Dataset Features

The scatter plot matrix shown in Fig. 5 gives a pair wise scatter plot of the features 'Mean', 'Variance', 'Standard Deviation', and 'Entropy', colored by the class (presence of a tumor). This type of plot can help identify relationships between features. However, due to the high density and overlapping points, it might be hard to see clear patterns.

The boxplot in Fig. 6 represents the distribution of selected features across the two classes. From these plots, we can observe that mean, variance and standard deviation are higher for images with a tumor (Class 1) than those without a tumor (Class 0). Entropy seems to have a slightly lower median value for images with a tumor (Class 1) than those without a tumor (Class 0). These visualizations confirm the observations we made earlier about the correlations between these features and the presence of a tumor. The box plots also allow us to observe potential outliers in the data, as points that are located outside the whiskers of the box plot.

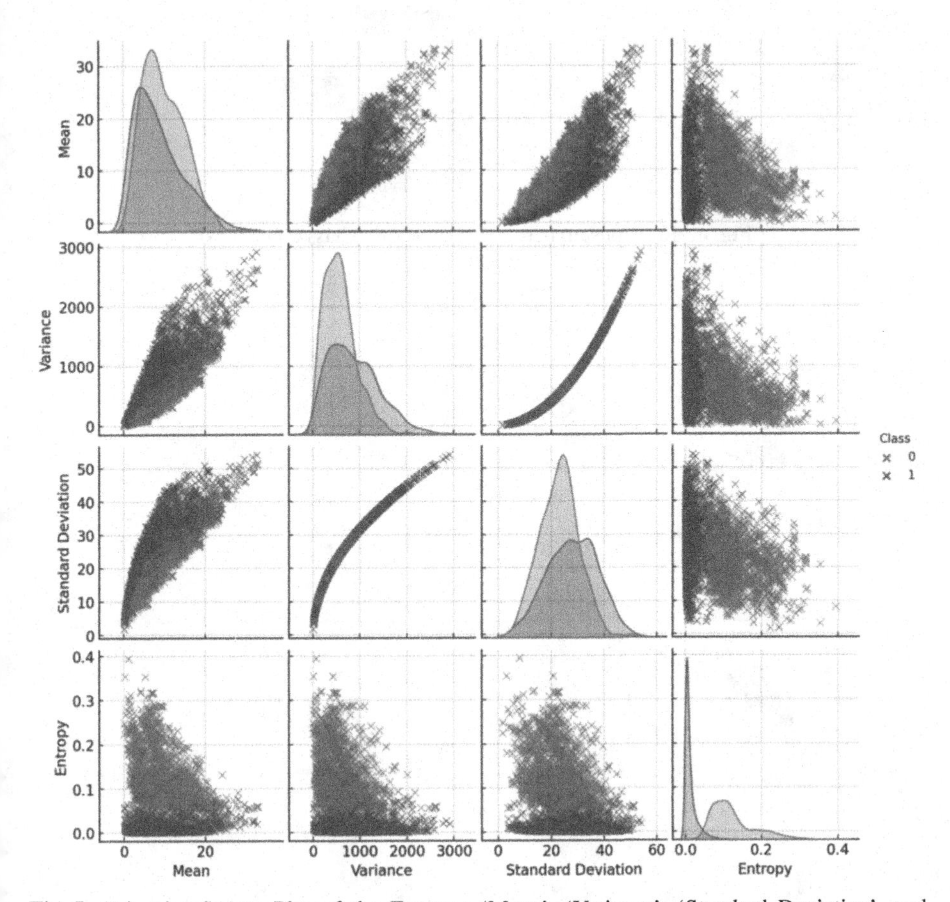

Fig. 5. Pair-wise Scatter Plot of the Features 'Mean', 'Variance', 'Standard Deviation', and 'Entropy'

The scatter plot matrices in Figs. 7 and 8 show pair-wise relationships between selected features (Mean, Variance, Standard Deviation, and Entropy) for images with a tumor (Class 1) and images without a tumor (Class 0) separately.

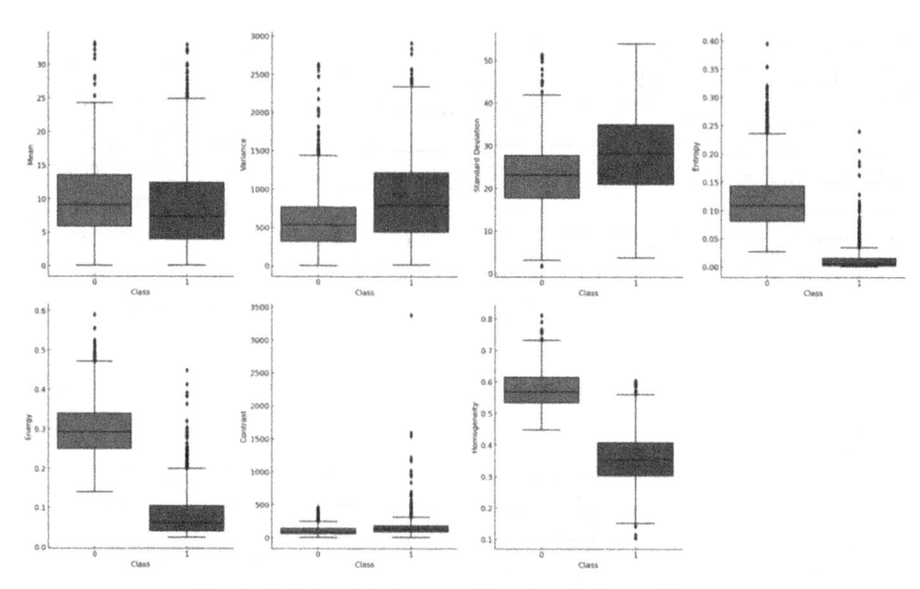

Fig. 6. Relationship between Class 1 and Class 0 Features

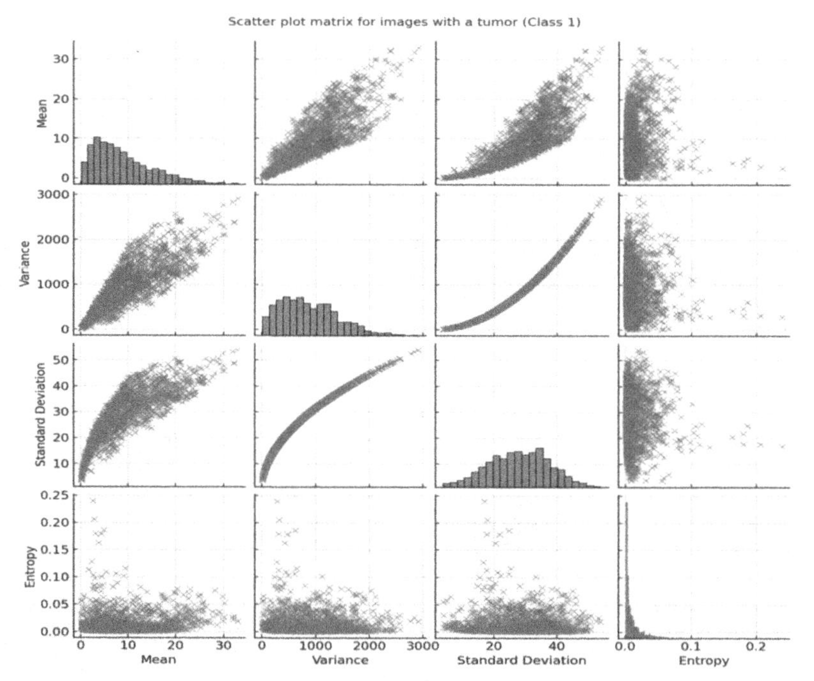

Fig. 7. Pair-wise Relationships between Selected Features (Mean, Variance, Standard Deviation, and Entropy) for Images with a tumor (Class 1)

It's interesting to note that the distributions and relationships between features appear to differ between the two classes. For instance, in images with a tumor, there seems to be a positive correlation between 'Mean' and 'Standard Deviation'. However, in images without a tumor, this correlation is less clear.

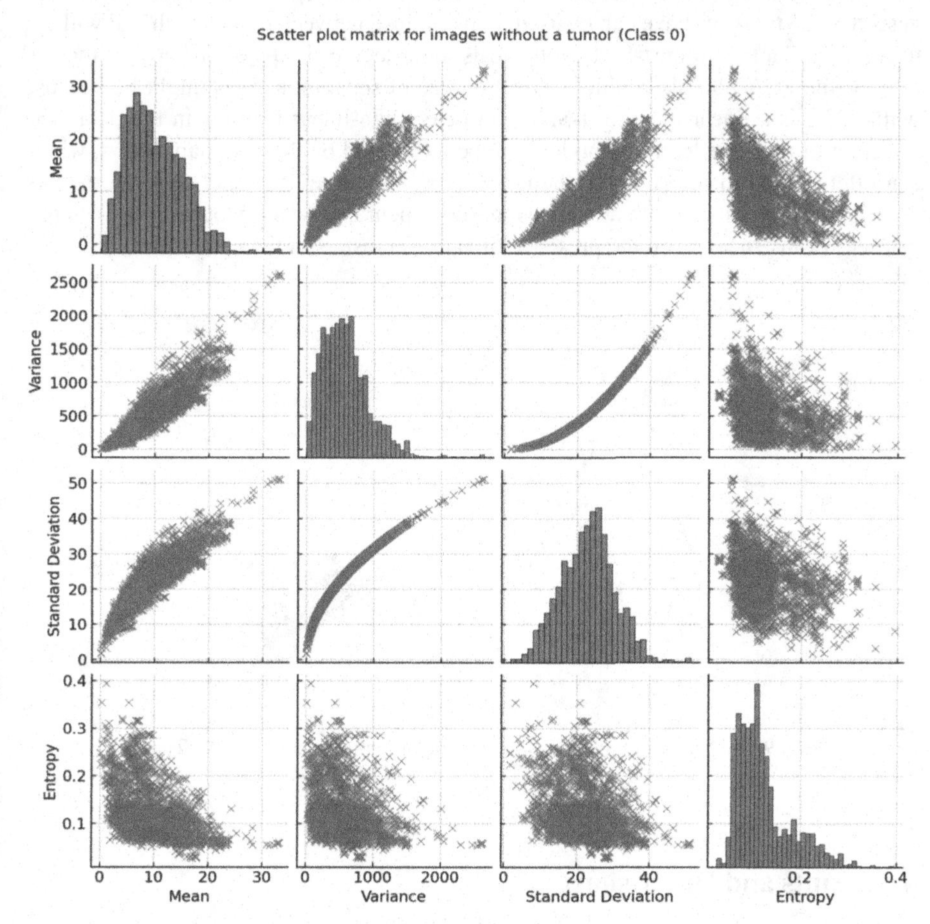

Fig. 8. Pair-wise Relationships between Selected Features (Mean, Variance, Standard Deviation, and Entropy) for Images with a tumor (Class 0)

The plots plotted in Fig. 9 below show the data projected onto a two-dimensional space using t-SNE and PCA, with different colors representing different classes (presence or absence of a tumor). These dimensionality reduction approaches aid in the visualization of high-dimensional data and may assist in identifying patterns or clusters. It is clear from the t-SNE plot that the two classes form separate clusters, which suggests that the traits we have can be used to distinguish between photos with and without tumors. Similar to how the t-SNE plot reveals some degree of separation between the two classes; the PCA plot also exhibits some degree of separation. This might be because while PCA is a linear method, t-SNE can detect non-linear features in the data. The selection of machine learning models can be influenced by these visualizations, which can offer insightful information about the data's structure. A model that can capture non-linear correlations, such as a decision tree or neural network, would do well on this dataset, for instance, given the presence of various clusters in the t-SNE plot.

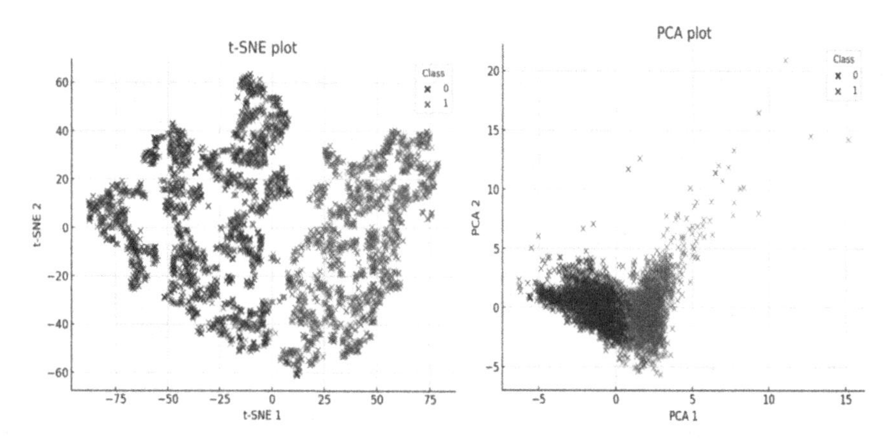

Fig. 9. t-SNE and PCA plots showing the presence or absence of a Tumor

4 Results and Discussions

The testing dataset included 334 occurrences of Class 1 (tumor) and 419 instances of Class 0 (no tumor). From Table 3, we can observe that the Gradient Boosting classifier and the Random Forest classifier have the highest accuracies (98.1% and 98.4%, respectively) among the models tested, making them suitable models for distinguishing between brain images with and without a tumor based on the given dataset. The Linear SVC, on the other hand, has the lowest accuracy (62.2%) among the tested models. The Random Forest model has the highest accuracy of approximately 98.4%, followed closely by the Gradient Boosting model with an accuracy of about 98.1%. The Decision Tree model has an accuracy of around 97.5%. The Logistic Regression and Gaussian Naive Bayes models also perform well, with accuracies of approximately 94.2% and 95.9%, respectively. The

K-Nearest Neighbors model has an accuracy of about 81.9%. The Linear SVC model has the lowest accuracy of approximately 62.2%. The plots shown in Figs. 10 and 11 provide a clear comparison of the performance of the different models on the test data.

Table 3. Results of Different Machine Learning Algorithms for Cancer Dataset

Algorithm	Findings	Accuracy
Gradient Boosting	True Negatives (Class 0): 416 False Positives (Class 1): 3 True Positives (Class 1): 323 False Negatives (Class 0): 11	98.1%
Linear SVC	True Negatives (Class 0): 419 False Positives (Class 1): 0 True Positives (Class 1): 49 False Negatives (Class 0): 285	62.2%
K-Nearest Neighbors	True Negatives (Class 0): 363 False Positives (Class 1): 56 True Positives (Class 1): 254 False Negatives (Class 0): 80	81.9%
Logistic Regression	True Negatives (Class 0): 407 False Positives (Class 1): 12 True Positives (Class 1): 302 False Negatives (Class 0): 32	94.2%
Gaussian Naive Bayes	True Negatives (Class 0): 410 False Positives (Class 1): 9 True Positives (Class 1): 312 False Negatives (Class 0): 22	95.9%
Random Forest	True Negatives (Class 0): 416 False Positives (Class 1): 3 True Positives (Class 1): 325 False Negatives (Class 0): 9	98.4%
Decision Tree	True Negatives (Class 0): 412 False Positives (Class 1): 7 True Positives (Class 1): 322 False Negatives (Class 0): 12	97.5%

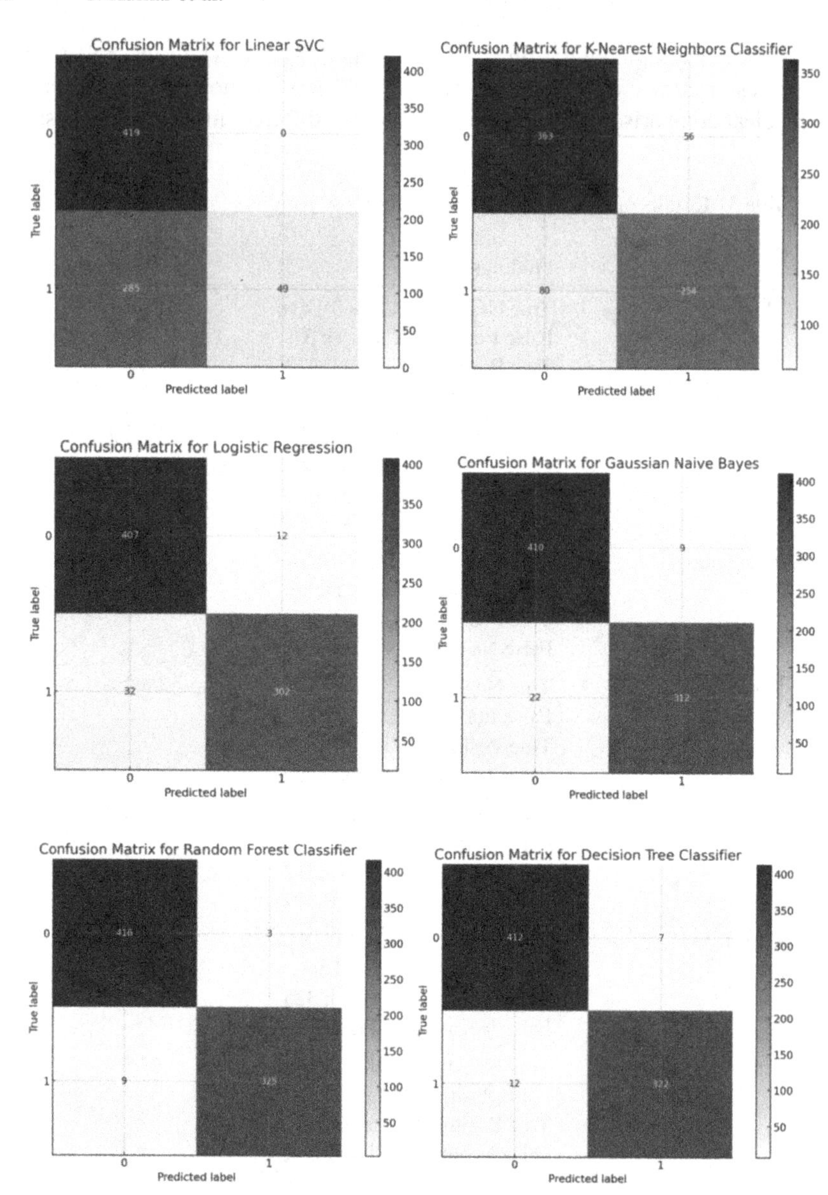

Fig. 10. (a–f): Confusion Matrix of Different Machine Learning Algorithms

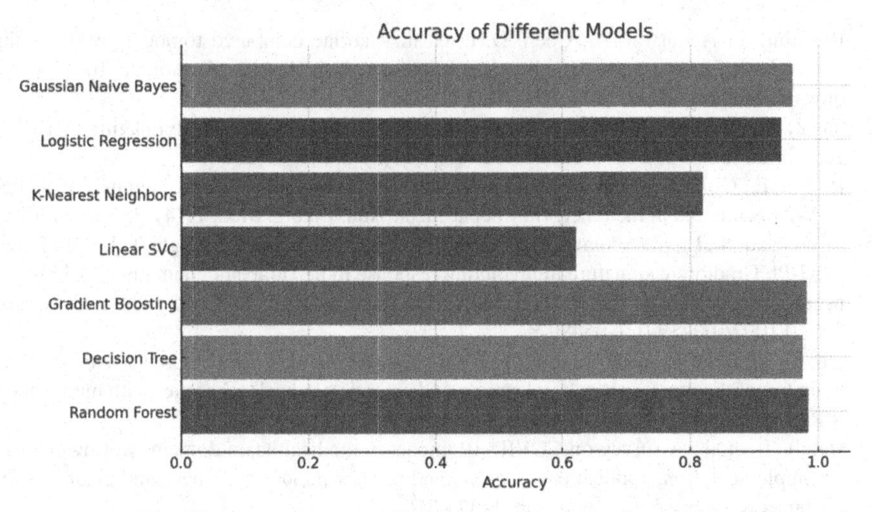

Fig. 11. Accuracy Plot of Different Machine Learning Models

5 Conclusion

In this article, the authors set out to pre-examine and classify data related to brain tumors using machine learning methods. The goal of the research is to determine whether machine learning algorithms could assist medical professionals in early brain tumor detection and classification. Initially, the authors thoroughly pre-examined the data on brain tumors, placing special emphasis on feature engineering and data preprocessing. We were able to enhance the dataset's quality and suitability for machine learning research with the aid of this phase. Only a few of the preprocessing steps included data cleaning, dealing with missing values, and scaling features. The dataset optimization we performed would provide the best possible foundation for our subsequent machine-learning models to make accurate predictions. Secondly, a range of machine learning techniques were employed, such as decision trees, support vector machines, neural networks, and random forests, to classify cases of brain cancers based on the dataset. The performance of models is positive in terms of accuracy, sensitivity, and specificity. These findings suggest that machine learning has the potential to be an effective tool for the early detection and classification of brain tumors, which can significantly aid medical professionals in making an early and accurate diagnosis. The promise of machine learning for brain cancer detection and classification is underlined in this study. The findings demonstrate the need to apply cutting-edge technologies to enhance the efficacy and accuracy of medical diagnostics.

References

1. Bahadur, N.B., Ray, A.K., Thethi, H.P.: Image analysis for MRI based brain tumor detection and feature extraction using biologically inspired BWT and SVM. Int. J. Biomed. Imaging **2017**, 12 (2017)

2. Rosmini, S., Aggarwal, A., Chen, D.H., et al.: Cardiac computed tomography in cardio-oncology: an update on recent clinical applications. Eur. Heart J. Cardiovasc. ImagingCardiovasc. Imaging **22**(4), 397–405 (2021)
3. Tayal, N., Tayal, S.: Advance computer analysis of magnetic resonance imaging (MRI) for early brain tumor detection. Int. J. Neurosci. **131**(6), 555–570 (2020)
4. Pivetta, E., Goffi, A., Tizzani, M., et al.: Lung Ultrasonography for the diagnosis of SARS-CoV-2 pneumonia in the emergency department. Ann. Emerg. Med. **77**(4), 385–394 (2021)
5. Whitehead, S., Li, S., Ademuyiwa, F.O., Wahl, R.L., Dehdashti, F., Shoghi, K.I.: Co-clinical FDG-PET radiomic signature in predicting response to neoadjuvant chemotherapy in triple-negative breast cancer. Eur. J. Nucl. Med. Mol. Imaging **49**(2), 550–562 (2021). https://doi.org/10.1007/s00259-021-05489-8
6. Ali, S., Li, J., Pei, Y., Khurram, R., Rehman, K., Mahmood, T.: A comprehensive survey on brain tumor diagnosis using deep learning and emerging hybrid techniques with multi-modal MR image. Arch. Comput. Methods Eng. **1**, 3 (2022)
7. Habib, T., et al.: A survey on COVID-19 impact in the healthcare domain: worldwide market implementation, applications, security and privacy issues, challenges and prospects. In: Complex & Intelligent Systems, pp. 1–32 (2022)
8. Hasan, A.M., Meziane, F., Aspin, R., Jalab, H.A.: Segmentation of brain tumors in MRI images using three-dimensional active contour without edge. Symmetry **8**(11), 132 (2016)
9. Ranjbarzadeh, R., BagherianKasgari, A., Ghoushchi, S.J., Anari, S., Naseri, M., Bendechache, M.: Brain tumor segmentation based on deep learning and an attention mechanism using MRI multi-modalities brain images. Sci. Rep. **11**(1), 10930 (2021)
10. Katta, A.K., Katta, R.B.: An automated brain tumor detection and classification from MRI images using machine learning techniques with IoT. Environ. Dev. Sustain. **2021**, 1–15 (2021)
11. Chithambaram, T., Perumal, K.: Automatic detection of brain tumor from magnetic resonance images (MRI) using ANN-based feature extraction. Int. J. Adv. Res. Eng. Technol. **12**(1), 109–118 (2021)
12. Thongam, M.M., Thongam, K.: Retracted article: computer-aided detection of brain tumor from magnetic resonance images using deep learning network. J. Ambient. Intell. Humaniz. Comput. **12**(7), 6911–6922 (2020)
13. Abbood, A.A., Shallal, Q.M., Fadhel, M.A., Shallal, Q.M.: Automated brain tumor classification using various deep learning models: a comparative study. Indones. J. Electr. Eng. Comput. Sci. **22**(1), 252 (2021)
14. Younis, A., Qiang, L., Nyatega, C.O., Adamu, M.J., Kawuwa, H.B.: Brain tumor analysis using deep learning and VGG-16 ensembling learning approaches. Appl. Sci. **12**, 7282 (2022)
15. Raza, A., et al.: A hybrid deep learning-based approach for brain tumor classification. Electronics **11**(7), 1146 (2022)
16. Alanazi, M.F., et al.: Brain tumor/mass classification framework using magnetic-resonance-imaging-based isolated and developed transfer deep-learning model. Sensors **22**(1), 372 (2022)
17. Khan, A.H., et al.: Intelligent model for brain tumor identification using deep learning. Appl. Comput. Intell. Soft Comput. **2022**, 1–10 (2022)
18. Zahid, U., et al.: BrainNet: optimal deep learning feature fusion for brain tumor classification. Comput. Intell. Neurosci. **2022**, 1 (2022)
19. Sharif, M.I., Khan, M.A., Alhussein, M., Aurangzeb, K., Raza, M.: A decision support system for multimodal brain tumor classification using deep learning. Complex Intell. Syst. **2021**, 1–14 (2021)
20. Sekhar, A., Biswas, S., Hazra, R., Sunaniya, A.K., Mukherjee, A., Yang, L.: Brain tumor classification using fine-tuned GoogLeNet features and machine learning algorithms: IoMT enabled CAD system. IEEE J. Biomed. Health Inform. **26**(3), 983–991 (2022)

21. Khan, S.A., SenthilVelan, S.: Application of exploratory data analysis to generate inferences on the occurrence of breast cancer using a sample dataset. In: 2020 International Conference on Intelligent Engineering and Management (ICIEM). IEEE (2020)
22. Sweetlin, E.J., Saudia, S.: Exploratory data analysis on breast cancer dataset about survivability and recurrence. In: 2021 3rd International Conference on Signal Processing and Communication (ICPSC). IEEE (2021)
23. Agaal, A., Essgaer, M.: An exploratory data analysis of breast cancer features in south of Libya. J. Pure Appl. Sci. **21**(4), 57–64 (2022)

Supplier Selection for Agriculture Industry Under Uncertainty: Machine Learning Based Sample Average Approximation Method

Ömer Faruk Yılmaz[1,2](✉) and Beren Gürsoy Yılmaz[1,2]

[1] Department of Industrial Engineering, Karadeniz Technical University, 61080 Trabzon, Turkey
omerfarukyimaz@ktu.edu.tr

[2] Department of Industrial and Systems Engineering, University of Florida, Gainesville 32608, USA

Abstract. This paper addresses the problem of supplier selection for a hazelnut producer in the agriculture industry under the uncertainty of product quality, with the objective of cost minimization. To mathematically represent the problem, a two-stage stochastic programming model is developed to make decisions both before and after the realization of uncertainty. The first stage involves supplier selection decisions, while the second stage determines the amount of product received from each supplier. After the hazelnuts are produced, if they do not meet quality standards, customers may return the products, causing significant customer satisfaction issues and increased costs. Product quality is assessed based on four features: (I) Aflatoxin, (II) Acidity, (III) Moisture, and (IV) Sifter. To address this problem within the context of intelligent agriculture, four different scenario reduction techniques are employed: (I) Sample Average Approximation (SAA), (II) SAA combined with Machine Learning-1 (SAA + ML1), (III) SAA + ML2, and (IV) SAA + ML3. Each method utilizes the K-means clustering algorithm with specific problem-related characteristics. Computational results indicate that when the SAA approach is combined with K-means clustering and scenarios are reduced based on clustering results, better outcomes are achieved in reducing costs associated with low-quality products. Specifically, higher numbers of suppliers and higher probabilities of generating scenarios from relevant clusters contribute to improved cost reductions.

Keywords: Intelligent Agriculture · Two-stage stochastic programming · K-means clustering · Sample average approximation

1 Introduction

Agriculture is one of the most vital industries, playing a strategic role in ensuring global food security. With the increasing world population, the demand for agri-food products is growing, necessitating a shift from traditional agricultural methods to smart agriculture practices [1]. In the agriculture industry, maintaining product quality while minimizing costs is crucial for producers. Hazelnut producers, in particular, face significant challenges due to inherent uncertainties in product quality stemming from various factors

S. Tiwari et al. (Eds.): AI4S 2024, CCIS 2243, pp. 184–195, 2025.
https://doi.org/10.1007/978-3-031-81369-6_15

such as weather conditions, soil quality, and farming practices. Ensuring that the final product meets quality standards is vital, not only for customer satisfaction but also for minimizing the costs associated with product returns. Consequently, supplier selection becomes a critical decision for hazelnut producers as it directly influences the quality of the final product [2].

The quality of hazelnuts can be assessed through specific features such as aflatoxin levels, acidity, moisture content, and the presence of sifter residues. These quality indicators are critical for ensuring that the final product meets both regulatory standards and consumer expectations [3]. The problem of supplier selection has been extensively studied over the past decade in the fields of operations research and supply chain management [3, 4]. Most of these studies apply multi-criteria decision-making (MCDM) approaches to identify the best supplier for the relevant industry [5, 6]. However, these methods heavily rely on the decision-makers' subjective judgments, which can lead to the selection of an unsuitable supplier. Therefore, it is essential to account for the uncertainty inherent in parameters such as the quantity of products received from each supplier [7]. Given that supplier-related parameters can fluctuate over time, handling inherent uncertainty with a stochastic modeling approach is both plausible and commonly employed [8, 9].

Two-stage stochastic programming has emerged as a powerful tool for decision-making under uncertainty [10, 11]. In this framework, decisions are made in two stages: the first stage involves decisions made before the uncertainty is realized, and the second stage involves recourse actions taken after the uncertainty is revealed. Scenario generation and reduction techniques are essential for the practical implementation of stochastic programming models. The Sample Average Approximation (SAA) method is a popular technique for approximating the expected value of a stochastic problem by averaging multiple sample scenarios [12]. Recent advancements in machine learning have introduced new methods for scenario generation and reduction, enhancing the efficiency and accuracy of these models [13]. K-means clustering, in particular, has been successfully applied in various scenario reduction contexts to group similar scenarios and reduce computational complexity [14, 15].

An examination of the existing academic literature on supplier selection reveals a gap in addressing the uncertainty inherent in the selection process. Therefore, this study aims to tackle the supplier selection problem for a hazelnut producer under quality uncertainty by using a two-stage stochastic programming model. Four different scenario reduction techniques are compared: SAA, SAA combined with three machine learning-based methods (SAA + ML-1, SAA + ML-2, SAA + ML-3). Each method employs K-means clustering with unique problem-specific characteristics. This study contributes to the literature by providing a novel application of stochastic programming and machine learning techniques in the agricultural supply chain, specifically focusing on quality-driven supplier selection. To fully benefit from the potential of the implemented methods, the study provides several important insights following the presentation of computational results. This research can be considered a steppingstone for intelligent agriculture operations suggested for implementation by enterprises within the context of Agriculture 4.0 [16].

The rest of the study is organized as follows: Sect. 2 is devoted to the problem description and optimization model. Section 3 includes computational experiments and analysis. Finally, Sect. 4 presents the conclusions.

2 Problem Description and Optimization Model

2.1 Problem Description

This study addresses the supplier selection problem by considering the uncertain nature of the quality characteristics of the products received from suppliers. Formally, the problem can be described as follows: a producer manufactures various hazelnut products and employs a set of suppliers, denoted by I, to meet a set of customer demands, denoted by C. . The quality of the hazelnuts is assessed with respect to four key metrics: (I) Aflatoxin, (II) Acidity, (III) Moisture, and (IV) Sifter presence.

Once the products are produced and delivered to customers, any failure to meet the quality requirements can result in customer requests for refunds. This not only incurs direct financial costs associated with product returns but also negatively impacts customer relationships with the producer. Consequently, managing the quality of the hazelnuts and minimizing the associated costs becomes a critical concern.

The primary objective is to select suppliers in such a way as to minimize the total cost, which includes both the costs of supplier selection and the costs associated with product returns. Given the uncertain nature of the product quality parameters, the problem is modeled mathematically using a two-stage stochastic programming approach. In this framework, decisions are made in two stages:

First Stage: Supplier selection decisions are made before the realization of uncertainty.

Second Stage: Decisions regarding the amount of products received from each supplier are made after the uncertainty has been realized.

The uncertainty in product quality is modeled through multiple scenarios, which represent possible realizations of quality metrics. By incorporating these scenarios, the model can effectively capture the variability in product quality and make more robust supplier selection decisions.

2.2 Two-Stage Stochastic Optimization Model

The following assumptions are considered while formulating the problem.

Assumptions

- The customers' demands are deterministic and known in advance.
- Quality thresholds and associated cost are known.

The following sets, parameters, and variables are employed in the model formulation.

Sets

I: Set of suppliers, indexed by i$\{i \in I\}$

C: Set of customers, indexed by c$\{c \in C\}$
S: Set of scenarios, indexed by s$\{s \in S\}$
K: Set of features, indexed by k $\{k \in K\}$

Parameters

c_i : Fixed cost to select supplier i
q_{isk} : Quality level of products with respect to criterion k received from supplier i for scenario s
t_k : Quality threshold for criterion k
g_k : Penalty cost for a unit of product for not satisfying criterion k
p_s : Probability associated with scenario s
Q_c: Demand of product
M: A large number

First-stage decision variables

x_i : If supplier i is selected to receive products, 1; otherwise, 0

Second-stage decision variables

y_{is} : The amount of products received from supplier i for scenario s

Objective function

$$z = \min \sum_{i \in I} c_i x_i + \underset{\xi}{E}\, \mathcal{Q}(x, \xi) \tag{1}$$

where,

$$E_\xi\, \mathcal{Q}(x, \xi) = \min \sum_{s \in S} p^s \left(\sum_{i \in I} \sum_{k \in K} \max(0, q_{isk} - t_k) g_k y_{is} \right) \tag{2}$$

subject to

$$\sum_{i \in I} y_{is} = \sum_{c \in C} Q_c;\ s \in S \tag{3}$$

$$y_{is} \le M x_i;\ i \in I,\ s \in S \tag{4}$$

$$0 \le y_{is}\ and\ x_i \in \{0, 1\} \tag{5}$$

The deterministic equivalent of the two-stage stochastic programming model is given above. Equation (1) represents the objective function, which includes both the first-stage and second-stage related costs: the total cost of selecting suppliers and the total expected cost of returned products from customers, as shown in Eq. (2). Equation (3) ensures that the total amount of demand is satisfied by receiving products from the suppliers under the relevant scenario. Equation (4) guarantees that if a supplier is not selected, no products can be received from it. Equation (5) shows the non-negativity and binary restrictions for the variables.

In this optimization model, the role of probabilities associated with each scenario is to aid in determining the expected recourse function, which is sensitive to these scenario probabilities. The uncertainty inherent in product quality characteristics is represented by these probabilities in the two-stage stochastic programming model. For the supplier selection process, several criteria are considered within the recourse function to account for multiple aspects of the products provided by suppliers. These probabilities are determined based on historical data from the relevant industry, thereby enabling the calculation of the expected recourse function value, which directly influences supplier selection decisions. Based on preliminary tests, the optimization model can solve problems with parameter settings up to $|I| = 100$ and $|S| = 300$ within reasonable CPU times. Since scenario reduction techniques are applied, the results are quite similar to those obtained with the settings used in the computational analysis.

3 Computational Experiments and Analysis

3.1 Case Study and Experimental Design

In this study, the problem addressed is derived from a real case study in the northeastern region of Turkey. In this case study, a producer of multiple hazelnut products has frequently encountered return orders from its customers. This issue highlights the need for the company, as well as the agriculture industry, to study the supplier selection problem from various perspectives. The values taken from the real case study for 12 suppliers, with respect to four features, are presented in Table 1.

Table 1. Feature values from 12 suppliers

Suppliers	Aflatoxin	Acidity	Moisture	Sifter
S-1	7.92	0.04	4.41	1.33
S-2	7.75	0.05	3.92	0.61
S-3	6.64	0.09	5.13	1.02
S-4	6.53	0.12	5.35	0.23
S-5	7.02	0.17	4.74	1.24
S-6	7.82	0.14	5.83	2.42
S-7	8.01	0.34	6.25	1.75
S-8	8.23	0.08	6.04	0.23
S-9	7.23	0.25	5.73	0.96
S-10	6.94	0.22	5.34	0.81
S-11	5.65	0.28	4.11	1.01
S-12	6.44	0.22	5.30	0.89
Unit	ug/kg	%/kg	%/kg	%/kg
(Acceptable upper bound levels)	≤10	1	6 (for natural)	≤10 (for 13–15 mm)

The scenarios for the two-stage stochastic programming model are generated based on the values for four features given in Table 1. Accordingly, aflatoxin values are assumed to be distributed uniformly as U[5–10]. Acidity values are assumed to be distributed uniformly as U[0–0.4]. Moisture values are assumed to be distributed uniformly as U[3–7]. Sifter values are assumed to be distributed uniformly as U[0–3].

To achieve generalized results, data are generated for different levels of the number of suppliers, specifically 10, 20, and 30. For 10 suppliers, two and three classes are considered. For 20 suppliers, three and four classes are considered. For 30 suppliers, four and five classes are considered. The following K-means clustering results are obtained with respect to the Aflatoxin, Acidity, and Moisture features for the scenario where the number of suppliers is equal to 10. The reason behind using these three features is based on expert opinions from the agricultural industry.

Table 2 demonstrates the clustering results for the scenario with 10 suppliers, with K = 2 and K = 3 representing the number of clusters. As observed from Table 2, the number of suppliers in each cluster is equal when K = 2. However, the number of suppliers in each cluster varies when for K = 3. This variation pattern is utilized in the scenario reduction methods.

Figures 1, 2, and 3 show the K-means clustering results with respect to two and three clusters for 10, 20, and 30 suppliers, respectively. When the number of suppliers is 10, two different clustering numbers (K = 2 and K = 3) are considered. Similarly, when the number of suppliers is 20, the considered number of clusters are K = 3 and K = 4. Moreover, when the number of suppliers is 30, the considered number of clusters are K = 4 and K = 5.

Table 2. K-means clustering for two and three clusters for number of suppliers equals to 10

7	Aflatoxin	Acidity	Moisture	Sifter	K = 2	K = 3
S-1	6.87	0.00	5.44	1.82	1	2
S-2	9.75	0.39	3.55	0.51	2	1
S-3	8.65	0.33	4.16	0.19	2	3
S-4	7.99	0.08	4.46	2.84	2	3
S-5	5.78	0.07	4.82	2.89	1	2
S-6	5.77	0.07	6.14	2.42	1	2
S-7	5.29	0.12	3.79	0.91	1	2
S-8	9.33	0.20	5.05	0.29	2	3
S-9	8.00	0.17	5.36	2.05	2	3
S-10	8.54	0.11	3.18	1.32	2	1

3.2 Sample Average Approximation Method

The SAA method is a widely used technique for solving stochastic optimization problems through Monte Carlo simulation [17, 18]. This approach approximates the expected objective function of a stochastic problem by using a sample average estimate derived

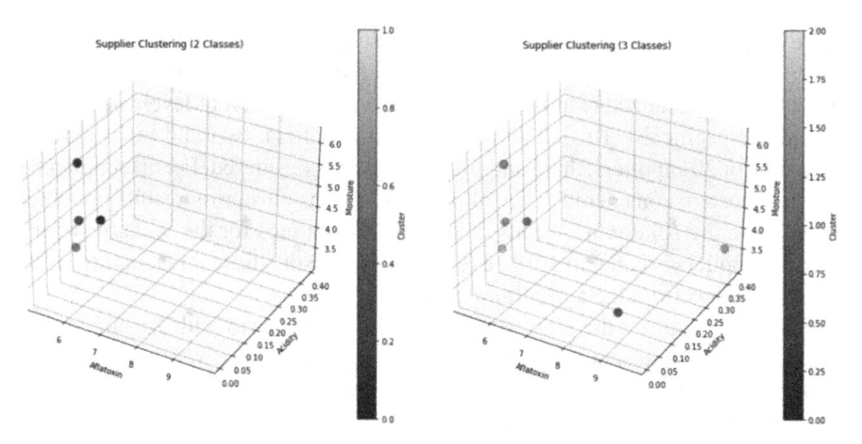

Fig. 1. K-means clustering results with respect to two and three clusters for 10 suppliers

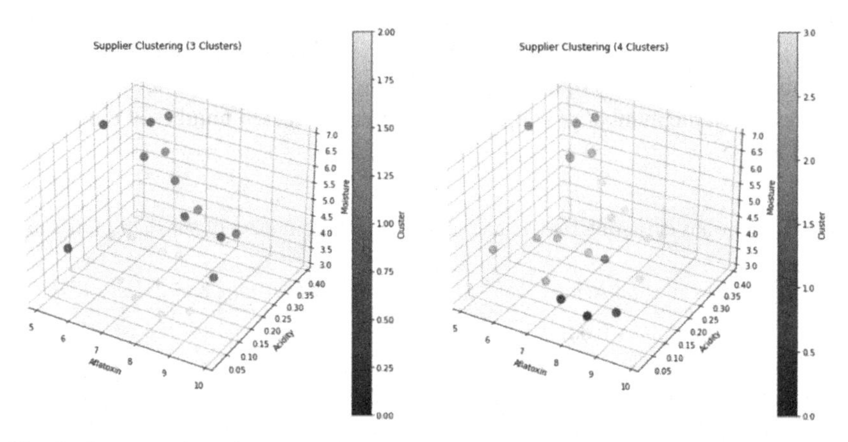

Fig. 2. K-means clustering results with respect to three and four clusters for 20 suppliers

from randomly generated samples [13]. Once this approximation is obtained, the resulting problem is treated as a deterministic optimization problem and is solved using conventional optimization techniques. To ensure robustness and accuracy, the process is repeated with different random samples. This repetition yields multiple candidate solutions and allows for the computation of statistical estimates of their optimality gaps. The sample average approximation problem given in Equation [6] converge to the original two-stage stochastic problem as the number of scenarios (N) goes infinity [17].

$$z_N = \min_{x \in X} c^T x + \frac{1}{N} \sum\nolimits_{n \in N} Q(x, \xi(\omega^n)) \tag{6}$$

3.3 Machine Learning Integrated SAA Methods

In this study, four different scenario reduction methods are compared to provide insights into the supplier selection problem. These methods are introduced as follows:

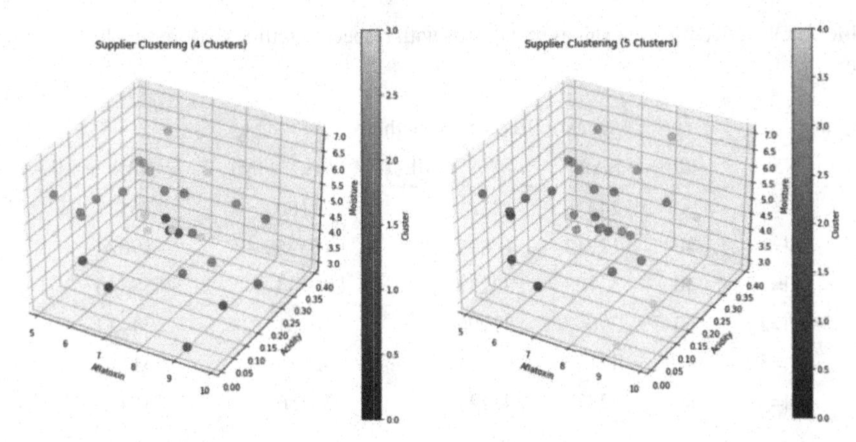

Fig. 3. K-means clustering results with respect to four and five clusters for 30 suppliers

- **SAA + ML-1:** This method utilizes the SAA approach combined with K-means clustering to reduce the number of scenarios. An equal number of scenarios are considered from each generated cluster.
- **SAA + ML-2:** This method also employs the SAA approach and K-means clustering for scenario reduction. However, the number of scenarios considered from each cluster is proportional to the percentage of suppliers in that cluster. The higher the number of suppliers in a cluster, the higher the probability of generating scenarios from that cluster.
- **SAA + ML-3:** Similar to the previous methods, this method uses the SAA approach and K-means clustering. The number of scenarios considered from each cluster is inversely proportional to the percentage of suppliers in the cluster. The fewer the suppliers in a cluster, the higher the probability of generating scenarios from that cluster.

3.4 Computational Results

In this section, the computational results obtained from each method with respect to combination of settings are provided in Table 3 without any prioritization among suppliers. In this table, for the settings, the K means number of clusters, NS stands for number of suppliers, |S| represents the number of scenarios.

According to the computational results, it is obvious that while the best results are achieved by SAA + ML-2, the worst results are obtained by SAA + ML-3 for all settings. The results achieved by SAA + ML-1 and SAA + ML-2 methods are similar to each other.

Figure 4 demonstrates the results in terms of the objective function obtained from each method. These results clearly indicate that when reducing scenarios, priority should be given to clusters with a higher number of suppliers. In other words, there is no need to consider an excessive number of scenarios from clusters with very few suppliers.

Furthermore, since clustering scenarios also involves grouping the cases caused by suppliers, products should be sourced from suppliers that are well-represented within a

Table 3. Objective function values of methods with respect to settings before supplier prioritization

Settings			Scenario Reduction Methods			
K	NS	|S|	SAA	SAA + ML-1	SAA + ML-2	SAA + ML-3
2	|I|=10	20	24418	23381	21180	26525
3	|I|=10	40	19913	18897	16525	19674
3	|I|=20	30	32798	33227	30144	38846
4	|I|=20	60	29963	31195	28861	34423
4	|I|=30	40	44182	45931	42285	51183
5	|I|=30	80	42371	41129	37756	47412

cluster. Reliable suppliers, who consistently provide products with stable metric values, should be prioritized in the selection process.

Table 4 presents the values after a prioritization scheme is applied to suppliers, considering the order of feature importance: 1-aflatoxin, 2-acidity, and 3-moisture. Figure 5 illustrates the results given in Table 4 in terms of the objective function value. According to the results, it can be concluded that the best outcomes are achieved by the SAA + ML-2 method, while the worst results are obtained by the SAA + ML-3 method across all settings. The results from the SAA + ML-1 and SAA + ML-2 methods are similar. The primary reason behind this fact is that in the SAA + ML-2 method, each cluster is constructed based on the percentage of suppliers in the relevant cluster. This approach allows for an even distribution of suppliers across clusters, where the suppliers are selected by applying the proposed optimization model. Thereby, this method performs better than others.

Furthermore, when comparing the results in Table 4 with those in Table 3, it is evident that the cost values are reduced after applying a prioritization scheme among suppliers, based on the importance of the features (1-aflatoxin, 2-acidity, and 3-moisture). This suggests that the importance of features also influences the cost coefficients used in the

Table 4. Objective function values of methods with respect to settings after supplier prioritization

Settings			Scenario Generation Methods			
NC	NS	|S|	SAA	SAA + ML-1	SAA + ML-2	SAA + ML-3
2	|I|=10	20	22158	21138	18863	24411
3	|I|=10	40	17713	17797	15471	18714
3	|I|=20	30	29326	31217	28309	35541
4	|I|=20	60	28815	28745	25973	32123
4	|I|=30	40	41135	40175	39911	46732
5	|I|=30	80	39714	38722	36142	41186

model. Additionally, it highlights that when suppliers are prioritized and selected from clusters with a higher number of scenarios, cost values can be reduced.

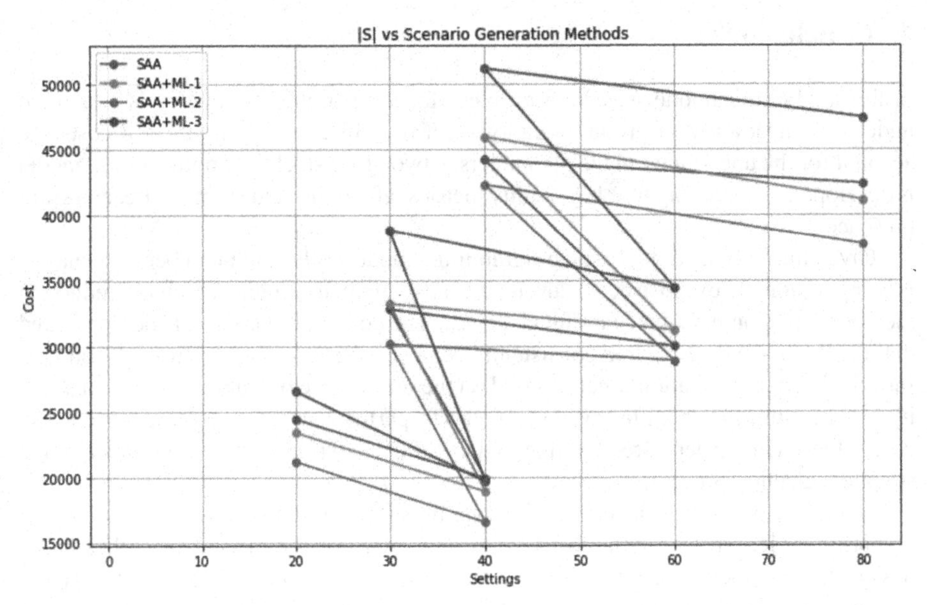

Fig. 4. Computational results for all methods before supplier prioritization

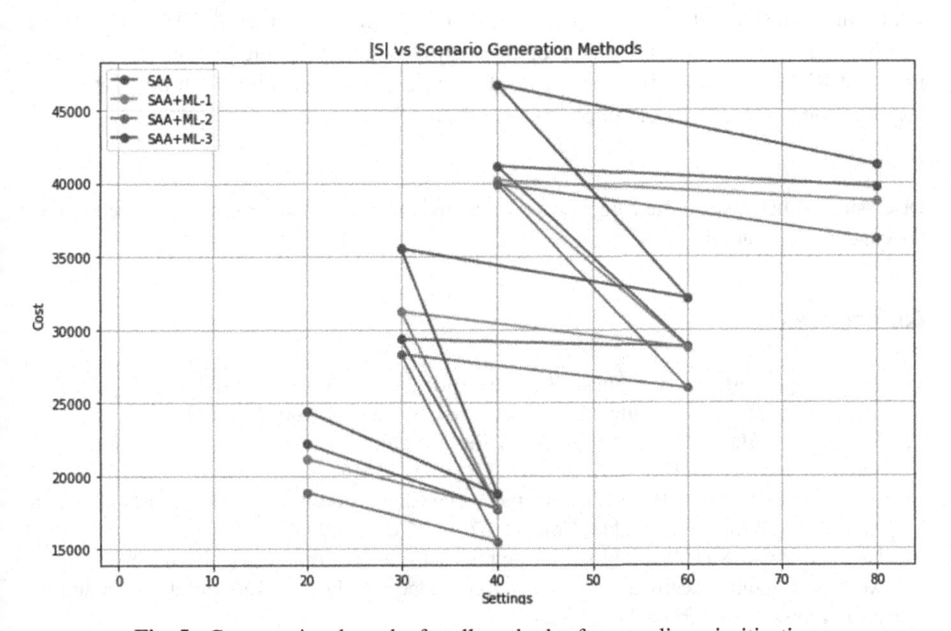

Fig. 5. Computational results for all methods after supplier prioritization

In both Figs. 5, a similar tendency is observed in terms of cost values with respect to |S| for all methods.

4 Conclusions

In this study, the supplier selection problem for the agriculture industry is addressed under the uncertainty of product quality, with the objective of minimizing costs. To account for the uncertainty in the parameters, a two-stage stochastic optimization model is developed, considering four key quality metrics: aflatoxin, acidity, moisture, and sifter presence.

Given that solving a stochastic programming model with a high number of scenarios is computationally expensive, we developed and compared combined scenario reduction methods based on machine learning and SAA. The computational results demonstrated that the SAA + ML-2 method consistently achieved the best results across all settings. Introducing a prioritization scheme based on the importance of features (aflatoxin, acidity, and moisture) further improved the method's performance, leading to a reduction in cost values. This underscores the significance of considering feature importance in the supplier selection process.

Our findings indicate that prioritizing suppliers from clusters with a higher number of scenarios can lead to substantial cost savings. This approach not only enhances the robustness of the supplier selection process but also aligns with the principles of intelligent agriculture, contributing to more efficient and sustainable agricultural practices.

For future research directions the following points can be considered. (I) The uncertainty inherent of other parameters such as demand can be considered. (II) The problem can be extended to cover all stages of the supply chain including customers. Last but not least, (III) other machine learning and deep learning algorithms can be adopted to improve the performance of supplier selection mechanism.

Disclosure of Interests. The authors have no competing interests to declare that are relevant to the content of this article.

References

1. Abbasi, R., Martinez, P., Ahmad, R.: The digitization of agricultural industry a systematic literature review on agriculture 4.0. Smart Agric. Technol. **2**, 100042 (2022)
2. Bozoğlu, M., Başer, U., Kilic Topuz, B., Alhas Eroglu, N.: An overview of hazelnut markets and policy in Turkey. KSU J. Agric. Nat. **22**(5), 733 (2019)
3. Wu, F., Khlangwiset, P.: Evaluating the technical feasibility of aflatoxin risk reduction strategies in Africa. Food Addit. Contam. **27**(5), 658–676 (2010)
4. Islam, S., Amin, S.H., Wardley, L.J.: A supplier selection and order allocation planning framework by integrating deep learning, principal component analysis, and optimization techniques. Expert Syst. Appl. **235**, 121121 (2024)
5. Abdulla, A., Baryannis, G.: A hybrid multi-criteria decision-making and machine learning approach for explainable supplier selection. Supply Chain Anal. **7**, 100074 (2024)

6. Sahoo, S.K., Goswami, S.S.: Green supplier selection using MCDM: a comprehensive review of recent studies. Spec. Eng. Manag. Sci. **2**(1), 1–16 (2024)
7. Islam, S., Amin, S.H., Wardley, L.J.: A supplier selection & order allocation planning framework by integrating deep learning, principal component analysis, and optimization techniques. Expert Syst. Appl. **235**, 121121 (2024)
8. Jana, C., Garg, H., Pal, M., Sarkar, B., Wei, G.: MABAC framework for logarithmic bipolar fuzzy multiple attribute group decision-making for supplier selection. Complex Intell. Syst. **10**(1), 273–288 (2024)
9. Yılmaz, Ö.F., Yeni, F.B., Yılmaz, B.G., Özçelik, G.: An optimization-based methodology equipped with lean tools to strengthen medical supply chain resilience during a pandemic: a case study from Turkey. Transp. Res. Part E Logist. Transp. Rev. **173**, 103089 (2023)
10. Yılmaz, Ö.F., Özçelik, G., Yeni, F.B.: Ensuring sustainability in the reverse supply chain in case of the ripple effect: a two-stage stochastic optimization model. J. Clean. Prod. **282**, 124548 (2021)
11. Grass, E., Fischer, K.: Two-stage stochastic programming in disaster management: a literature survey. Surv. Oper. Res. Manag. Sci. **21**(2), 85–100 (2016)
12. Mitrai, I., Palys, M.J., Daoutidis, P.: A two-stage stochastic programming approach for the design of renewable ammonia supply chain networks. Processes **12**(2), 325 (2024)
13. Kim, S., Pasupathy, R., Henderson, S.G.: A guide to sample average approximation. Handbook of simulation optimization, pp. 207–243. Springer, Heidelberg (2015)
14. Bertsimas, D., Kallus, N.: From predictive to prescriptive analytics. Manag. Sci. **66**(3), 1025–1044 (2020)
15. Hu, S., Dong, Z.S., Dai, R.: A machine learning based sample average approximation for supplier selection with option contract in humanitarian relief. Transp. Res. Part E Logist. Transp. Rev. **186**, 103531 (2024)
16. Dvorkin, Y., Wang, Y., Pandzic, H., Kirschen, D.: Comparison of scenario reduction techniques for the stochastic unit commitment. In: 2014 IEEE PES General Meeting| Conference & Exposition, pp. 1–5. IEEE, Maryland (2014)
17. Verweij, B., Ahmed, S., Kleywegt, A.J., Nemhauser, G., Shapiro, A.: The sample average approximation method applied to stochastic routing problems: a computational study. Comput. Optim. Appl. **24**, 289–333 (2003)
18. Kleywegt, A.J., Shapiro, A., Homem-de-Mello, T.: The sample average approximation method for stochastic discrete optimization. SIAM J. Optim.Optim. **12**(2), 479–502 (2002)

Enhancing Cyclone Preparedness: Deep Learning Methods with INSAT-3D Satellite Imagery

K. Aditya Shastry(✉), B. S. Aneesh, M. P. Chinmay, C. Gowtham Patel, and G. N. Shashank

Nitte Meenakshi Institute of Technology, Bengaluru 560064, India
adityashastry.k@nmit.ac.in

Abstract. Cyclones are enormous storms that bring heavy rain and strong winds. The INSAT 3D satellite accurately tracks cyclones and their progression. The objective of our work is to measure the severity of the cyclone using the generated sequence of photographs. Estimating cyclone strength is essential for timely alerts, risk analysis, emergency management, and understanding cyclone behaviour. It helps with planning, resource allocation and minimising the effects on people's lives and infrastructure. Current cyclone intensity detection systems lack complexity and dynamics, leading to inaccurate predictions and ineffective reaction measures. Improved strategies for detecting cyclone intensity can improve catastrophe preparedness and response in cyclone-prone locations. This work is centered on the study of cyclone intensity prediction using INSAT-3D satellite images. INSAT stands for "Indian National Satellite System". Utilizing historical data and atmospheric features captured by INSAT-3D, we employ deep learning models to capture complex temporal correlations. These architectures excel in anticipating cyclone strength, facilitating timely alerts and effective emergency management. The comparative analysis demonstrates that our proposed Convolutional Neural Network model performed better over existing methods. By embracing cyclone dynamics, we provide precise intensity assessments, enhancing catastrophe preparedness and response.

Keywords: Recurrent neural networks · long short-term memory networks · Cyclone estimation · Satellite imagery

1 Introduction

Accurately estimating the strength of tropical cyclones (TC) is vital for efficient reaction to their devastating impact on people and property. Satellite remote sensing facilitates several methods for TC monitoring. IR satellite imagery is commonly used for both traditional and automated Dvorak methods. Other algorithms, such as DAVT(deviation angle variance technique) and CNN(Convolution neural network), are still under investigation. Satellite data is commonly used to estimate TC intensity using microwave sounder-based or consensus methodologies [3].

S. Tiwari et al. (Eds.): AI4S 2024, CCIS 2243, pp. 196–207, 2025.
https://doi.org/10.1007/978-3-031-81369-6_16

Deep learning methods process satellite photos and meteorological data to estimate cyclone intensity. Models often use CNNs for spatial patterns and RNNs for sequential dependencies. Trained models provide predictions for disaster preparedness and response. Deep learning estimates cyclone strength by leveraging pre-trained models on huge datasets, addressing data scarcity [1]. Performance indicators such as MAE(mean absolute error) and RMSE(root mean square error) evaluate forecasting accuracy. Real-time integration of cyclone data collection systems is crucial for accurate intensity estimates. Deep learning methods for estimating cyclone strength have challenges due to limited and low-quality data. Challenges include capturing complex cyclone behaviour driven by marine and atmospheric elements, along with obtaining scarce and inconsistent data.

Data augmentation strategies, including as rotation, scaling, and flipping, improve the generalizability and effectiveness of deep learning techniques using limited labelled cyclone data. Transfer learning leverages pre-trained models, commonly learned on sizable imagery datasets such as ImageNet, as initializations for cyclone intensity estimate. Deep learning techniques can potentially enhance efficiency. With less training data by using pretrained models' learned representations and feature extraction [2].

2 Related Work

This segment examines certain works which were performed in estimating wind velocity from satellite photos and current portals that offer real-time coastal storm information. Additionally, we offer background knowledge on CNNs, the building blocks of our deep learning model to increase the accuracy of the model.

Terrestrial cyclone intensity is commonly estimated using satellite photography by use of the Dvorak Technique, which was created by meteorologist Vernon Dvorak in the 1970s. The method's primary component is a close assessment of the cloud patterns connected to these storms. A centrally concentrated area of deep convection close to the centre of the cyclone is called the Central Dense Overcast, or CDO. This forms an important factor to be considered. An important factor in influencing the storm's severity is the CDO's presence and organisation. Furthermore, if an eye filled with clouds appears, the method considers this as a sign of a stronger tropical storm. The Dvorak Technique uses alphanumeric classifications, including T-numbers and CI numbers, to measure the storm's strength. Weather scientists can monitor changes in intensity over time with the help of these classifications. Based on this method, the Dvorak Satellite Intensity Estimates (DSI) offer a numerical representation of the storm's intensity, providing useful data for operational forecasting and study. For remotely determining the severity of tropical cyclones, especially when direct observations could be difficult, the Dvorak Technique is still an essential resource. Meteorological agencies throughout the realm use it extensively to improve their comprehension and estimation of these complex weather systems [5].

In the domain of deep learning, CNNs have shown themselves to be useful for cyclone intensity detection. Satellite imagery is essential to the research of cyclones, and CNNs are particularly good at extracting longitudinal data. CNNs are able to extract complex spatial patterns from photos. These networks can automatically learn and recognise pertinent aspects, such as cloud formations, eyewall structures, and other spatial characteristics associated with cyclone severity, in the case of cyclone satellite imagery. It is frequently necessary to identify hierarchical features at several scales to detect cyclone strength. Convolutional layers with different-sized filters are employed by CNNs to help the network learn hierarchical feature representations. This is especially helpful when examining cyclones with intricate and varied structures. It is possible to fine-tune pre-trained CNN models for cyclone intensity identification, particularly those that were trained on huge image datasets. Transfer learning lessens the need for a substantial volume of training records by enabling the model to apply expertise from general picture identification tasks to the unique properties pertinent to cyclones. CNNs are capable of processing multispectral satellite imagery and combining data from many modalities, like visible and infrared light spectrums. Through this fusion, the network is better able to identify details related to variations in cyclone intensity that would not be visible in individual spectral bands. CNNs can analyse enormous amounts of data efficiently, that is beneficial for real-time cyclone monitoring. CNN-based models help meteorologists make more accurate and timely predictions by continually analysing incoming satellite photos and providing fast information on changes in cyclone intensity. CNNs have the capability to be built for end-to-end learning, which enables them to directly map input satellite photos to forecasts of cyclone severity. Using a comprehensive approach can streamline the modelling process and enhance the model's capacity to represent intricate correlations between image characteristics and cyclone strength [5].

The DAVT was illustrated and utilized to the "North Atlantic" by authors in [6] and to the "North Pacific" by [8, 9]. This technique measures the balance of humid storms in IR satellite imageries. To that brightness, it conducts a directed gradients statistical evaluation. The degree of divergence of the gradients signal from the ideal circumferential plane indicates the gradients signal's congruence degree. The storm is measured using the variability of this departure angle. The technique has two main drawbacks: (i) it needs photos with appropriately labeled tropical cyclone centers; and (ii) it uses several models and appropriate settings for tropical cyclones in various locations, which reduces its application globally.

The capacity of RNNs to model temporal sequences makes them beneficial for analysing time-dependent meteorological data related to cyclones, which is why RNNs are useful. Long-term dependencies, multivariate time series, and variable-length input sequences are among the many tasks that RNNs are excellent at. To explain the intricate nonlinear dynamics inherent in cyclone formation, they are successful in predicting cyclone strength by considering the sequential context of previous data. LSTM and GRU are examples of advanced RNN changes which are generally utilized in deep learning for cyclone analysis despite their limitations in mitigating problems connected to long-term dependency [6].

The capacity of Long Short-Term Memory (LSTM) networks to recognise and maintain long-term correlations in successive meteorological information makes them effective for deep learning applications such as cyclone intensity detection. In addition to being adept at managing irregular time intervals and multivariate time series analysis, LSTMs are particularly well-suited for modelling the dynamic temporal patterns associated with cyclone development. Adaptable to nonlinear dynamics and less sensitive to sequence length, LSTMs automatically identify pertinent information and capture complex atmospheric patterns, making them effective instruments for precisely forecasting changes in cyclone power [6].

3 Proposed Work

CNNs, drawing inspiration from the organization of the visual cortex, employ compact overlapping neurons to process 2D visual data, facilitating tolerance to image translation. Figure 1 shows the overall design of the proposed work.

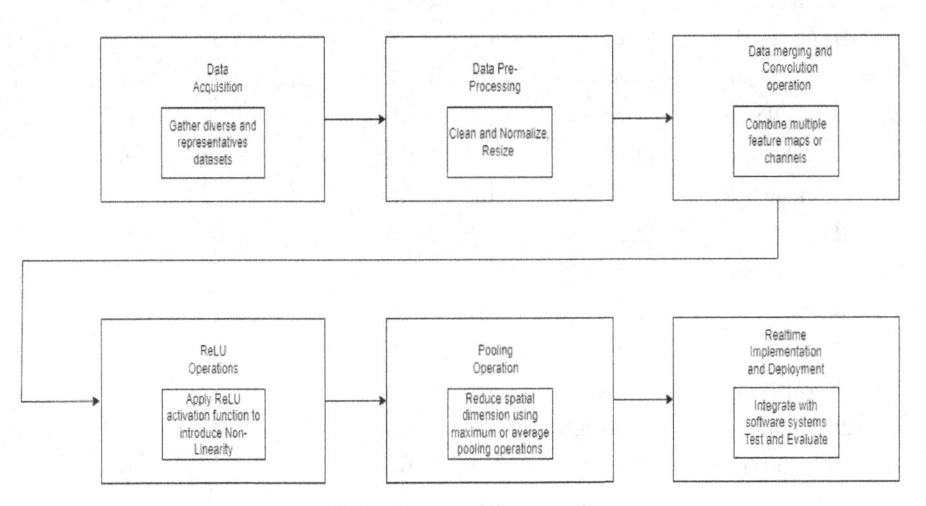

Fig. 1. Proposed framework

Table 1 shows the configuration of the CNN model proposed in this work for cyclone intensity estimation.

Table 1. Configuration of our convolutional neural network

Layer	Input Shape	Output Shape	Parameter Shape	Parameters
Input	–	(256,256,3)	–	0
Conv2D (1st)	(256,256,3)	(256, 256, 256)	(3, 3, 3, 256)	7168
Batch Normalization	(256, 256, 256)	(256, 256, 256)	(256,)	256
Conv2D (2nd)	(256, 256, 256)	(256, 256, 256)	(3, 3, 256, 256)	590080
Batch Normalization	(256, 256, 256)	(256, 256, 256)	(256,)	256
MaxPooling2D	(256, 256, 256)	(128, 128, 256)	–	0
Conv2D (3rd)	(128, 128, 256)	(128, 128, 256)	(3, 3, 256, 256)	590080

(continued)

Table 1. *(continued)*

Layer	Input Shape	Output Shape	Parameter Shape	Parameters
Batch Normalization	(128, 128, 256)	(128, 128, 256)	(256,)	256
Conv2D (4th)	(128, 128, 256)	(128, 128, 128)	(3, 3, 256, 128)	295040
Batch Normalization	(128, 128, 128)	(128, 128, 128)	(128,)	128
MaxPooling2D	(128, 128, 128)	(64, 64, 128)	–	0
Conv2D (5th)	(64, 64, 128)	(64, 64, 128)	(3, 3, 128, 128)	147584
Batch Normalization	(64, 64, 128)	(64, 64, 128)	(128,)	128
Conv2D (6th)	(64, 64, 128)	(64, 64, 64)	(3, 3, 128, 64)	73792
BatchNormalization	(64, 64, 64)	(64, 64, 64)	(64,)	64
MaxPooling2D	(64, 64, 64)	(32, 32, 64)	–	0
Conv2D (7th)	(32, 32, 64)	(32, 32, 64)	(3, 3, 64, 64)	36928
Batch Normalization	(32, 32, 64)	(32, 32, 64)	(64,)	64
Conv2D (8th)	(32, 32, 64)	(32, 32, 32)	(3, 3, 64, 32)	18464
Batch Normalization	(32, 32, 32)	(32, 32, 32)	(32,)	32
MaxPooling2D	(32, 32, 32)	(16, 16, 32)	–	0
Conv2D (9th)	(16, 16, 32)	(16, 16, 32)	(3, 3, 32, 32)	9248
Batch Normalization	(16, 16, 32)	(16, 16, 32)	(32,)	32
Conv2D (10th)	(16, 16, 32)	(16, 16, 16)	(3, 3, 32, 16)	4624
Batch Normalization	(16, 16, 16)	(16, 16, 16)	(16,)	16
Flatten	(16, 16, 16)	(1024,)	–	0
Dense (Output Layer)	(1024,)	(1,)	(1024, 1)	1025

3.1 Deep Convolutional Neural Network

CNNs effectively process 2D visual data, drawing inspiration from the organization of the animal visual cortex. Neural-inspired models, like LeCun et al.'s, are prevalent. Receptive fields, overlapping sub-regions in the visual field, are obtained through small neurons across CNN layers, allowing tolerance to image translation. CNNs, variants of multilayer perceptrons, comprise convolutional and wholly linked layers. The number of feature maps in a convolution layer equals its kernels. These feature maps feed into wholly linked tiers for classification, and back-propagation optimizes the network. Shared parameters with sparse connectivity address overfitting. The pooling layer, often using max pooling, follows the convolutional layer, reducing features and enhancing robustness. Backpropagation methods, exemplified by Kelley, are extensively employed for pattern identification. Deep architectures excel in high-level feature abstraction. Various deep learning models achieve state-of-the-art results in computer vision tasks. Training deep CNNs is challenging; GPUs are employed for faster training. Alex et al. (2012) used GPUs to win the ImageNet contest, followed by Zeiler and Fergus's (2013) improved model [4, 7].

3.2 Proposed Deep CNN Layers and Architecture

Our deep CNN architecture is meticulously designed to process 2D visual data efficiently, exhibiting a robust framework for complex tasks. The network begins with an input layer of dimensions (256, 256, 3), signifying a 256 × 256 pixel representation through 3 color channels [3, 7]. This initial layer serves as the gateway for visual information to traverse through subsequent layers, initiating the intricate process of feature extraction. Convolutional layers play a pivotal role in our architecture, exemplified by the 1st Conv2D layer. Operating on the input data, these layers employ small 3 × 3 kernels to systematically scan and extract essential features from the images. The insertion of batch normalization at each convolutional stage ensures stability during the training process, contributing to the network's overall reliability [2, 4].

Strategic placement of max pooling layers is the basis of our architecture. Positioned after convolutional layers, max pooling serves to downsample spatial dimensions, reducing the computational load and enhancing the network's ability to learn hierarchical representations. This progressive reduction in feature map dimensions is a deliberate design choice, enabling the network to capture intricate details and patterns in the visual data [1, 9]. Continuing through the architecture, additional Conv2D layers and batch normalization contribute to the network's depth and complexity. The combination of convolutional operations and batch normalization guarantees that the network can adapt to the intricacies of the input data while maintaining stability throughout the learning procedure. The architecture culminates in a wholly linked layer, generating an output shape of table (1). This layer is responsible for synthesizing the learned features and making predictions. The usage of shared parameters across the network mitigates the risk of overfitting, enhancing the model's generalization capabilities [2, 3].

Furthermore, our deep CNN architecture leverages GPU acceleration to tackle the computational demands associated with training large-scale neural networks. This optimization facilitates faster training times and enables the model to handle extensive datasets effectively. In summary, our deep CNN architecture is a meticulously crafted framework for 2D visual data processing. It combines convolutional and pooling layers with strategic batch normalization to extract and learn intricate features from input images. The architecture's depth, coupled with shared parameters and GPU acceleration, positions it as an effective means for resolving complicated visual tasks.

3.3 Optimization

Larger convolution filter sizes are employed for processing larger inputs, gradually decreasing in size for higher layers. The count of filters in each layer is established by the network's capacity and task complexity. The choice of pooling filter size is also crucial, as larger filters significantly reduce parameters but may result in information loss. Striking a balance between filter size and information preservation is essential to prevent overfitting, making the selection process challenging. It is crucial to tailor the granularity of these choices built on the dataset and task complexity.

Regularization is a method to counter overfitting by penalizing higher-order attributes for a smoother learning curve. In our experiments, the model obtained at approximately 90% validation accuracy is selected using early stopping. To prevent overfitting and enhance performance, we employ the dropout method with a general dropout rate of p = 0.5 in our model. Dropout helps avoid underfitting and contributes to overall model robustness [1, 4].

4 Experimental Results

In this segment, we examine the dataset, training and testing, visualization of features, and performance analysis.

4.1 Cyclone Dataset

The Infrared and Raw Cyclone Imagery dataset is an extensive compilation of satellite images gathered by the Indian National Satellite System (INSAT) 3D satellite. Encompassing the time frame from 2012 to 2021, this dataset is a valuable repository for investigating cyclonic phenomena and atmospheric conditions across the Indian Ocean region. The dataset comprises two primary types of imagery:

- Infrared Imagery: These are high-resolution images capturing thermal patterns within the atmosphere. By detecting variations in temperature, these images aid in identifying and monitoring cyclones. Infrared imagery provides crucial insights into the temperature differences and thermal gradients within the atmosphere, enabling researchers to track and analyze the development, movement, and intensity of cyclonic systems.
- Raw Cyclone Imagery: This subset of satellite images focuses specifically on cyclonic activities. These images are in their original, unprocessed form, offering a detailed view of cyclones as they evolve and intensify. Researchers can utilize this raw data to conduct in-depth analyses of the physical adaptations, patterns, and dynamics of cyclones over time. By leveraging the INSAT3D dataset, researchers and meteorologists can gain a comprehensive understanding of cyclone dynamics, study atmospheric conditions, and potentially enhance forecasting models. The dataset's temporal scope from 2012 to 2021 provides a significant span of data, facilitating longitudinal studies and trend analysis related to cyclones in the Indian Ocean region [7, 9].

Figure 2 depicts the sample images of the INSAT3D dataset used in our research work.

4.2 Training and Test

Our research leveraged the Infrared and Raw Cyclone Imagery dataset, an extensive compilation of satellite images acquired from the Indian National Satellite System (INSAT) 3D satellite covering the temporal span of 2012 to 2021. This dataset proved invaluable for both training and testing phases, offering a diverse range of atmospheric conditions and cyclonic phenomena. The dataset comprises two primary types of imagery: Infrared Imagery and Raw Cyclone Imagery. Infrared Imagery consists of high-resolution images capturing thermal patterns within the atmosphere, aiding in the identification and monitoring of cyclones. These imageries offer perceptions into temperature variations and

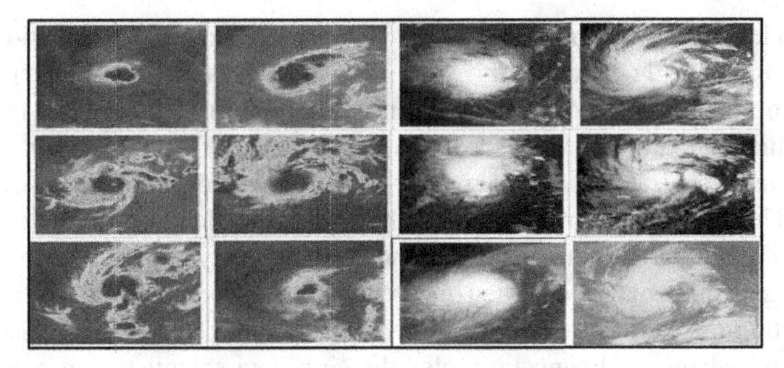

Fig. 2. Sample images from the Cyclone dataset

thermal gradients, enabling the tracking of cyclonic system development, movement, and intensity.

Simultaneously, the Raw Cyclone Imagery subset focuses on unprocessed satellite images that offer a detailed view of cyclones in their original state. This subset facilitates indepth analyses of structural changes, patterns, and dynamics associated with cyclones. For the training phase, the INSAT3D dataset ensured a robust and diverse foundation for model training [10]. The richness of data over the specified time frame allowed our model to learn effectively and generalize across various atmospheric conditions. The training process involved exposing the model to both Infrared Imagery and Raw Cyclone Imagery, enhancing its ability to recognize and understand cyclonic phenomena.

Subsequently, the testing phase evaluated our model's functioning on a distinct set of information from the similar Infrared and Raw Cyclone Imagery dataset. This dataset served as a reliable source for assessing the model's predictive capabilities on unseen data. The testing dataset, spanning the same years as the training data, included both Infrared Imagery and Raw Cyclone Imagery subsets. The Infrared Imagery subset in the testing dataset provided a benchmark for evaluating the model's accuracy in identifying and monitoring cyclones based on thermal patterns [2, 4]. Meanwhile, the Raw Cyclone Imagery subset permitted us to assess the model's capability to analyze unprocessed images depicting cyclonic activities. By employing the INSAT3D dataset for both training and testing, our research aimed to ensure the model's robustness, generalization, and reliability in predicting cyclonic phenomena across the Indian Ocean region. The comprehensive nature of the dataset contributed to the effectiveness of our model in understanding and responding to diverse atmospheric conditions.

4.3 Performance Analysis

The evaluation of the proposed CNN model was conducted using a combination of visualizations, including a scatter plot comparing actual versus predicted values and a confusion matrix. These plots deliver effective perceptions into the model's predictive accuracy and its ability to correctly classify cyclone images into different classes (Class 0 and Class 1). In the scatter plot, every point signifies a data instance, with the x-axis indicating the actual class values and the y-axis representing the corresponding predicted values generated by the model. The scatter plot provides a visual understanding of how well the model aligns with the ground truth, [1, 9] showcasing the distribution of points along the diagonal and helping identify any patterns or discrepancies in the predictions.

Following the scatter plot, the code generates a confusion matrix, a fundamental tool for evaluating classification models. The confusion matrix breaks down the model's predictions into four categories: true positives, true negatives, false positives, and false negatives. This matrix is particularly insightful in understanding the model's performance, highlighting areas where it excels and areas that may require improvement. The heatmap visualization of the confusion matrix enhances interpretability, with color-coded cells indicating the frequency of each type of prediction. Overall, these visualizations serve as comprehensive diagnostic tools, allowing data scientists and researchers to gain a nuanced understanding of the model's strengths and weaknesses. The inclusion of both a scatter plot and a confusion matrix contributes to a holistic assessment of the model's predictive capabilities, facilitating informed decisions for model refinement and optimization.

Table 2 shows the comparative evaluation of our designed custom CNN model with existing models with regards to the R2 metric.

Table 2. Comparison of results

Models	R^2
Sequential-CNN	0.9471
Proposed-CNN	0.9799
Xception	0.9428
ResNet-18	0.9167

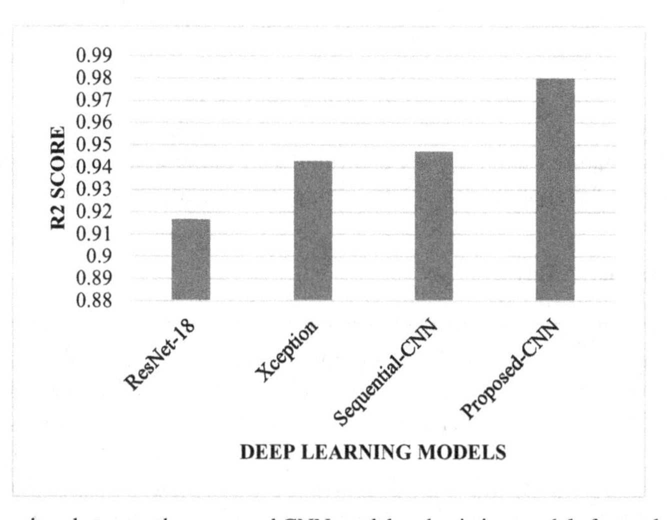

Fig. 3. Comparison between the proposed CNN model and existing models for cyclone intensity estimation on INSAT 3D dataset.

Figure 3 depicts the comparison between the existing and the proposed CNN model for cyclone intensity estimation on INSAT3D dataset.

As can be observed, the proposed CNN model performed with R2 improvements ranging from 3.28% to 6.32%.

5 Discussion

Cyclones are enormous storms that bring heavy rain and strong winds. The INSAT-3D satellite accurately tracks cyclones and their progression. The objective of our work is to measure the severity of cyclones using the generated sequence of photographs. Estimating cyclone strength is essential for timely alerts, risk analysis, emergency management, and understanding cyclone behavior. It helps with planning, resource allocation, and minimizing the effects on people's lives and infrastructure. Current cyclone intensity detection systems lack complexity and dynamics, leading to inaccurate predictions and ineffective reaction measures. Improved strategies for detecting cyclone intensity can improve catastrophe preparedness and response in cyclone-prone locations.

This work is centered on the study of cyclone intensity prediction using INSAT-3D satellite images. INSAT stands for "Indian National Satellite System." Utilizing historical data and atmospheric features captured by INSAT-3D, we employ deep learning models to capture complex temporal correlations. These architectures excel in anticipating cyclone strength, facilitating timely alerts and effective emergency management. The comparative analysis demonstrates that our proposed Convolutional Neural Network (CNN) model performed better than existing methods. By embracing cyclone dynamics, we provide precise intensity assessments, enhancing catastrophe preparedness and response.

While our study primarily focuses on Infrared (IR) imagery obtained from the INSAT-3D satellite, this approach may limit the model's effectiveness in utilizing other potentially valuable data sources, such as microwave imagery and additional meteorological parameters. Incorporating these alternative data sources could enhance the model's ability to capture different aspects of cyclone behavior, leading to more robust predictions. Future research should explore the integration of these diverse data types to improve prediction accuracy and reliability.

Additionally, the current model lacks uncertainty quantification in its predictions, which is a critical aspect for reliable decision-making, especially in emergency management scenarios. Incorporating uncertainty quantification into the model would allow for a more comprehensive risk assessment, enabling better preparedness and response strategies. Addressing these limitations will be a priority for future research, with the goal of developing a more comprehensive and reliable system for cyclone intensity prediction.

6 Conclusion and Future Scope

Our work introduces a real-time, end-to-end deep learning system for estimating the cyclone intensity using deep learning. The system proposes a newly developed CNN model, specifically designed for objectively estimating cyclone wind speed based solely on satellite images. Moreover, it is noted that the time spent developing the algorithm is significantly less than generating a comprehensive, consistent training dataset of imagery and related wind velocities. Deploying the machine learning model is a challenging process requiring multiple repetitions with updated training data and model configurations. From a software engineering standpoint, the model is considered as a source program and appropriately versioned. Ultimately our research classifies the given cyclone images as normal, medium and high(parameters for classification of cyclone intensities) and helps in catastrophe alertness.

The research's future endeavors should include exploring the advantages of multimodal data fusion beyond Infrared (IR) imagery, a primary focus in the current study. Incorporating data from diverse sources such as microwave imagery and meteorological parameters has the potential to significantly improve cyclone intensity detection. A vital research avenue involves developing a deep learning framework capable of effectively integrating these varied data modalities. Additionally, investigating transfer learning and generalization is crucial to assessing how well models trained on specific regions or time periods can adapt to others. Understanding the models' versatility and applicability across different geographical locations is vital for broader utility.

Addressing the lack of uncertainty quantification in deep learning predictions is another priority. Future research should concentrate on developing methods to quantify uncertainty, enhancing decision-making reliability in cyclone intensity detection. Furthermore, recognizing the dynamic nature of cyclones, characterized by rapidly changing intensities, calls for the establishment of models that can adapt in real-time. Incorporating temporal dynamics into the model architecture and training process is vital for improving the model's accuracy in predicting evolving cyclonic conditions. These future research directions aim to advance cyclone intensity detection by incorporating diverse data sources, quantifying uncertainty, and capturing the dynamic nature of cyclones.

References

1. Pradhan, R., Aygun, R.S., Maskey, M., Ramachandran, R., Cecil, D.J.: On Tropical Cyclone Intensity Estimation Using a Deep Convolutional Neural Network, vol. 27, no. 2 (2018)
2. Devaraj, J., Ganesan, S., Elavarasan, R.M., Subramaniam, U.: A novel deep learning based model for tropical intensity estimation and post-disaster management of hurricanes. Appl. Sci. **11**(9), 4129 (2021)
3. Kar, C., Kumar, A., Banerjee, S.: Tropical cyclone intensity detection by geometric features of cyclone images and multilayer perceptron. SN Appl. Sci. **1**, 1–7 (2019)
4. Lee, J., Im, J., Cha, D.-H., Park, H., Sim, S.: Tropical cyclone intensity estimation using multi-dimensional convolutional neural networks from geostationary satellite data. Remote Sens. **12**(1), 108 (2019)
5. Maskey, M., et al.: Deepti: deep-learning-based tropical cyclone intensity estimation system. IEEE J. Sel. Top. Appl. Earth Obs. Remote Sens. **13**, 4271–4281 (2020)
6. Meng, F., Xie, P., Li, Y., Sun, H., Xu, D., Song, T.: Tropical cyclone size estimation using deep convolutional neural network. In: 2021 IEEE International Geoscience and Remote Sensing Symposium IGARSS, pp. 8472–8475. IEEE (2021)
7. Tan, J., Yang, Q., Junjun, H., Huang, Q., Chen, S.: Tropical cyclone intensity estimation using himawari-8 satellite cloud products and deep learning. Remote Sens. **14**(4), 812 (2022)
8. Tian, W., Zhou, X., Huang, W., Zhang, Y., Zhang, P., Hao, S.: Tropical cyclone intensity estimation using multi- dimensional convolutional neural network from multichannel satellite imagery. IEEE Geosci. Remote Sens. Lett. **19**, 1–5 (2021)

9. Wang, C., Zheng, G., Li, X., Qing, X., Liu, B., Zhang, J.: Tropical cyclone intensity estimation from geostationary satellite imagery using deep convolutional neural networks. IEEE Trans. Geosci. Remote Sens. **60**, 1–16 (2021)
10. Patil, H.: Cyclone prediction from remote sensing images using hybrid deep learning approach based on AlexNet. Int. J. Image Graph. Signal Process. **1**, 96–107 (2024). https://doi.org/10.5815/ijigsp.2024.01.07

Machine Learning Detection of Depression Indicators in Online Communication

Kanchapogu Naga Raju and Sachi Nandan Mohanty(✉)

School of Computer Science and Engineering (SCOPE), VIT-AP University, Amaravati, Andhra Pradesh, India

nagaraju.23phd7048@vitap.ac.in, sachinandan09@gmail.com

Abstract. This research investigates the predictive modeling of depression through language patterns in online discourse, specifically analyzing user-generated content from Reddit's "depression" subreddit. Leveraging a dataset acquired via the Pushshift API, spanning January 1, 2015, to January 2, 2023, we applied advanced natural language processing (NLP) techniques—including Tokenization, StopWords removal, Lemmatization, and Word2Vec vectorization—to preprocess the text data. The study's core objective was to determine the efficacy of text-based features in predicting depression-related posts. To this end, a variety of machine learning models were trained and evaluated, including Logistic Regression, Random Forest, LSTM, Bi-LSTM, and Bi-LSTM-RNN. Our findings are quantified through precision, recall, and F1 scores, with a particular emphasis on model performance in accurately classifying depressive content. The comparative analysis reveals significant insights into the language of depression, providing a framework for future research and potential real-world applications in mental health monitoring and support systems.

Keywords: Predictive Analytics · Depression · Comparative Analysis · Deep Learning (DL) · Machine Learning (ML)

1 Introduction

In the digital age, the veil of anonymity provided by online platforms has allowed individuals to express personal struggles with unprecedented candor. This phenomenon has given rise to vast repositories of textual data, reflecting the collective mental health state of diverse populations. Among the most pressing of these conditions is depression, a pervasive disorder characterized by persistent sadness and a lack of interest or pleasure in previously rewarding or enjoyable activities. The ubiquity of depression and its profound impact on individuals and societies necessitates innovative approaches for early identification and intervention [1]. Traditional clinical diagnostics are invaluable; however, they are often constrained by resource availability and individuals' willingness to seek professional help. Consequently, there is a critical need to augment these approaches with additional tools that leverage the spontaneous and unsolicited expressions of affect found in online environments. ML, a subset of AI, has emerged as a powerful tool to harness this wealth of data. By analyzing patterns within large-scale textual data, researchers can glean insights into the linguistic cues associated with depression.

S. Tiwari et al. (Eds.): AI4S 2024, CCIS 2243, pp. 208–224, 2025.
https://doi.org/10.1007/978-3-031-81369-6_17

The subtleties of language—word choice, sentence structure, and thematic elements—can reveal underlying emotional states and cognitive processes [2]. This predictive capability is not only a scientific pursuit but also holds potential for real-world applications in mental health monitoring, offering a complimentary avenue for supporting individuals grappling with depression. This study seeks to contribute to this burgeoning field of inquiry by systematically examining how machine learning models can discern and predict depressive content within online textual exchanges. By bridging computational techniques with psychological insights, the research aims to illuminate the intersection of language and mental health, opening pathways for enhanced understanding and support for those affected by depression. Deep learning, a sophisticated branch of machine learning, has revolutionized the way data is interpreted and utilized across numerous domains. Its significance lies in its ability to learn and model complex, non-linear relationships within large sets of data. In the field of mental health, deep learning algorithms have become invaluable, offering new vistas for understanding and predicting depression through computational analysis of textual data.

At the heart of deep learning's transformative power is its neural network architecture, inspired by the neural structures of the human brain, allowing for intricate pattern detection. DL models, particularly those involving LSTM networks and their bidirectional counterparts (Bi-LSTMs), excel in processing sequential data, capturing the temporal dynamics of language as expressed in writing. This capability is paramount when analyzing online communications, which often comprise nuanced and sequential narrative threads. The utility of deep learning in detecting linguistic patterns indicative of depression cannot be overstated [3]. Unlike traditional machine learning models that require manual feature extraction, deep learning models autonomously identify relevant features—a process that is especially beneficial for unstructured data like text. This automation facilitates a more profound and nuanced analysis, enabling the models to capture subtleties in language usage that may signal depressive states [4]. In the context of mental health, the deployment of deep learning offers a dual advantage: it supports the identification of early warning signs of depression, and it provides a non-invasive method to monitor public mental health trends [5]. As such, deep learning not only serves as a tool for individual diagnosis and intervention but also has the potential to inform public health strategies and policy-making. The primary objective of this research is to harness deep learning's robust analytical capabilities to identify linguistic indicators of depression in written text. We aim to develop and validate a predictive model that not only accurately classifies text as indicative of depressive sentiment but also enhances our understanding of how language reflects mental health states. By focusing on the subtleties of language—such as semantic patterns, syntax, and sentiment—expressed in online platforms, this study seeks to bridge the gap between qualitative insights and quantitative analysis.

In pursuit of this goal, we will assess the performance of various deep learning architectures, including LSTM, Bi-LSTM, and Bi-LSTM-RNN models, against traditional machine learning benchmarks. The study strives to determine the extent to which deep learning models can outperform conventional algorithms in the context of natural language processing (NLP) tasks related to mental health. This work is propelled by the conviction that early and accurate detection of depression-related communication can lead to timely intervention and support. As such, the overarching aim is to contribute a novel approach to the proactive monitoring and analysis of mental health discourse, potentially influencing the development of digital tools for mental health professionals and providing invaluable support to those in need.

2 Related Work

[6] In a study focused on early detection of depression through Arabic speech analysis, a hybrid model integrating a CNN with a SVM was proposed. This model, tested on a benchmark dataset of 200 Arabic speech samples, demonstrated superior accuracy rates of 90.0% and 91.60% in training and testing phases respectively, compared to individual performances of RNN and CNNs. The hybrid approach not only outperformed these models but also showed lower false positive and negative rates, highlighting its potential utility for clinical practitioners in diagnosing depression from speech.

[7] Research exploring the use of DL to detect depression from audio and video cues highlights the technological strides in recognizing this mental disorder. By reviewing various databases and DL methods, this study provides an overview of the extraction of depression indicators from speech and facial expressions. It emphasizes the need for enhanced automatic depression diagnosis methods amidst the increasing psychological pressures of modern life and discusses the future challenges and opportunities in the field.

[8] The paper introduces "DepNet," a framework utilizing sequences of facial images from videos to diagnose depression, leveraging multiple pretrained models for feature representation. This method, tested on the AVEC2013 and AVEC2014 databases, achieved notably lower root mean-square errors compared to other video-based depression recognition methods. The integration of a Feature Aggregation Module allows for a comprehensive analysis of high-level characteristics, aiding in the accurate assessment of depression severity. [9] Addressing the challenge of depression diagnosis through social media, this research proposes a model based on LSTM and RNN to analyze textual data. Achieving an accuracy of 99.0%, the model excels in detecting depressive signs from text, surpassing other frequency-based deep learning approaches. The study underscores the potential of using advanced ML techniques to identify early signs of depression, which could significantly impact mental health management on social media platforms.

[10] This study investigates the potential of speech-based diagnostic tools for depression by utilizing a multi-information joint decision algorithm on the DAIC-WOZ dataset. By incorporating preprocessing steps such as noise reduction and feature extraction through OpenSmile, the model achieves an impressive 87% accuracy in diagnosing depression. Principal component analysis further refines the feature set, enhancing the convolutional neural network's effectiveness in this clinical application, indicating the viability of voice analysis as a tool for diagnosing depression.

[11] Utilizing structural magnetic resonance imaging (sMRI), this research identifies key brain regions linked to depression symptoms via deep learning. The study focuses on differentiating five depression symptom phenotypes, achieving notable predictive accuracy particularly for Anxiety and Suicidality. Deep learning models pinpointed

the anterior cingulate and orbitofrontal cortex as critical areas for symptom prediction, suggesting these regions' significant roles in targeted treatments for late-life depression (LLD).

[12] By analyzing polysomnographic data with deep learning, this study advances our understanding of the intersection between sleep disturbances and psychiatric conditions such as anxiety and depression. Employing the Xception model trained on RGB-transformed sleep data, the approach reached an accuracy of nearly 97.88% in detecting depression-related patterns among patients. The study underscores the potential of machine learning in psychiatry, particularly in leveraging sleep study data for diagnosing mental health conditions.

[13] Addressing the prevalence of depression and its underestimation, this paper explores the application of NLP to detect depression from social media posts, specifically on Twitter. Using techniques like TF-IDF and Word2Vec for feature generation, and evaluating various machine learning models, the study found that the Extra Trees classifier performed best, achieving an accuracy of 84.83%. This method demonstrates the effectiveness of analyzing sentiment in social media posts for early detection of depression, offering a promising tool for mental health monitoring.

[14] This research utilized deep learning on a large dataset of medical claims data from Estonia (2018–2022) to demonstrate the effectiveness of AI-based digital decision support systems (DDSS) in predicting depression. Using a combination of non-sequential and sequential models, including a novel Att-GRU-decay model, the study achieved high performance with an AUC of 0.990 and specificity of 0.999. The results underline the significance of temporal data properties and the potential of DDSS in a general practitioner setting, suggesting improvements in healthcare cost efficiency and quality of care for depression screening. [15] In addressing the challenge of early depression detection on social media, this paper presents a novel approach using a Deep Learning MDHAN. The model, applied to preprocessed Twitter data, significantly outperformed traditional models like CNN and SVM, achieving a remarkable 99.86% accuracy. This was facilitated by innovative feature selection methods, including Adaptive Particle and Grey Wolf optimization, highlighting the potential of advanced deep learning techniques in accurate and efficient depression detection from social media content.

[16] Focusing on the underexplored area of depression detection in Arabic social media content, this study leverages a Bi-LSTM model with an attention mechanism to analyze approximately 6000 tweets. The model's attention mechanism aids in identifying significant word weights that contribute to depression detection, achieving an accuracy of 0.83%. This approach underscores the complexities of the Arabic language and the potential of deep learning to address gaps in mental health diagnostics in non-English speaking populations.

[17] This paper introduces an innovative approach to depression detection utilizing embedded LSTM and NLP. The system, designed as a brain health mapping tool, integrates psychological traits to analyze textual interactions and detect emotional states. The performance of this system, evidenced by superior accuracy and F1 scores compared to existing methods, showcases the potential of ICT in personalized mental health monitoring, providing a novel framework for tracking and supporting users' mental health through their digital behaviors.

[18] This paper explores the utilization of EEG features for aiding the diagnosis of depression, employing a Bi-LSTM network that considers the time series impact on data accuracy. The study focuses on the recognition of depression through EEG data analyzed with a Bi-LSTM model based on event-related potentials (ERP). The enhanced Bi-LSTM model shows promising results, achieving an accuracy of 80.6% in identifying and classifying depression, which demonstrates its potential in improving diagnostic precision using EEG data.

[19] This work presents an attention-based LSTM model that makes use of frame-level properties of sad speech rather than the more conventional statistical features. The goal of this study is to improve the diagnosis of depression. Increasing the depth of feature representation is accomplished by the model's use of attention layers across both the time dimension and the feature dimension. It is noteworthy that this attentive LSTM model attained an average accuracy of 90.2% when it was tested on the DAIC-WOZ database. This model outperformed both regular LSTM and localized attention variations, demonstrating that it is effective in capturing the intricacies of depressive speech. [20] This research presents a CNN-LSTM-based multimodal model designed to analyze depression and anxiety using facial expression features and scale scores (SAS and SDS). By integrating CNNs and LSTMs, the model effectively processes and fuses multimodal data from video frames and scale results, achieving a high diagnostic accuracy of 94.6%. This approach demonstrates the utility of leveraging video-based facial expressions alongside standardized psychological scales to enhance the accuracy of initial screenings for depression and anxiety disorders.

[21] Addressing the rapid pre-diagnosis of MDD, this study utilizes a LSTM model enhanced with an attention mechanism to analyze facial features from interview videos. Incorporating techniques such as label smoothing and model explanation through integrated gradients, the model delivers robust diagnostic metrics, with an F1- score of 88.89% and accuracy of 91.67%. The findings emphasize the model's ability to capture critical facial expressions and behaviors indicative of depression, offering insights into effective features for depression identification and confirming its utility in clinical assessments. [17] This paper presents a case study on the classification of cauliflower diseases using deep learning techniques. The research demonstrates how CNNs can be effectively utilized to detect and classify plant diseases, enhancing agricultural sustainability and management practices.

[22] The study develops predictive machine learning models for water quality assessment. Presented at IC-ICN 2023, it outlines the utilization of various algorithms to accurately predict contaminants in water sources, which is crucial for ensuring environmental health and public safety.

[23] This research utilizes DL assisted FLAIR segmentation combined with genomic analysis to explore the heterogeneity of lower-grade gliomas. The approach enhances the understanding and treatment strategies for these brain tumors by integrating advanced imaging analysis with genetic data. [24] The paper focuses on the recognition and prediction of potato leaf diseases using convolutional neural networks. By improving the accuracy of disease diagnosis in plants, the study contributes to more effective agricultural interventions and disease management.

[25] This research addresses cardiovascular disease forecasting in the Bangladeshi population through an all-inclusive approach using both machine learning and deep learning methods. It emphasizes the development of predictive models that could assist in early diagnosis and better management of cardiovascular health. [26] The paper explores the identification and categorization of yellow rust infection in wheat using deep learning techniques. By applying advanced image recognition methods, the study aids in the accurate detection of this common wheat disease, supporting effective crop management strategies.

[27] This research makes use of deep learning and machine learning for the purpose of identifying and diagnosing rice leaf illnesses in Bangladesh in real time. The research focuses on agricultural applications. The implementation of these technologies in real-world contexts highlights the potential of these technologies to improve disease diagnosis and agricultural productivity. [28] Through the use of intelligent picture identification, this study investigates the application of machine learning and deep learning for the purpose of deciphering microorganisms. This article examines the current techniques, challenges, and breakthroughs in the discipline of microbiology, stressing the tremendous impact that these technologies have had in the sciences associated with microbiology.

[29] The study investigates advanced deep learning models for corn leaf disease classification through a field study in Bangladesh. By applying these models directly in the field, the research shows how deep learning can be effectively utilized for plant disease identification, supporting agricultural decision-making. [30] This research examines the application of ML and DL techniques for skin cancer detection. By analyzing various skin image data, the study aims to improve diagnostic accuracy and provide insights into more effective skin cancer screening methods.

Reference	Strengths	Threats	Accuracy (%)
[6]	Hybrid CNN and SVM model improves accuracy using Arabic speech analysis	Limited dataset size and language-specific model may not generalize	91.60
[7]	Comprehensive review of DL methods in detecting depression from audio and video	May not address all variability in real-world scenarios due to lab conditions	87.00
[8]	Utilizes video sequences with advanced feature aggregation for improved performance	Performance may degrade with lower quality video data	89.00
[9]	High accuracy using LSTM and RNN for text-based depression detection	Dependence on text data quality and specific language characteristics	99.0
[10]	Speech-based diagnosis using detailed speech signal processing and machine learning	Speech processing techniques may not capture all nuances of emotional expression	87.00

(*continued*)

(*continued*)

Reference	Strengths	Threats	Accuracy (%)
[11]	Uses deep learning on sMRI to predict depression symptoms, identifying key brain regions	Method may not be applicable in all clinical settings due to sMRI availability	88.00
[12]	High accuracy in detecting anxiety and depression using deep learning on sleep data	Relies on accurate sleep data collection and might not generalize to different populations	96.88
[13]	Analyzes sentiment from Twitter posts using NLP to detect depression	Social media data can be noisy and not always indicative of clinical depression	84.83
[14]	Demonstrates the use of large-scale medical data with deep learning for depression prediction	May overlook individual variations in depression symptoms	99.00
[15]	Introduces a novel hierarchical attention network for depression detection on Twitter	High accuracy might not translate to other social media platforms or text formats	99.86
[16]	Focus on underexplored Arabic social media content using Bi-LSTM with attention mechanism	Limited by the complexities of the Arabic language and smaller dataset	83.00
[17]	Proposes an embedded LSTM scheme for depression detection via ICT	Effectiveness may depend heavily on user engagement with the system	90.00
[18]	Employs Bi-LSTM for EEG-based depression diagnosis considering temporal dynamics	Specific to EEG data, which might not be readily available in all clinical settings	80.6
[19]	Uses attention layers in LSTM for nuanced capture of depressive speech features	Focused on depressive speech; may require high-quality audio data	90.2
[20]	Combines CNN and LSTM in a multimodal approach to diagnose depression from video and scales	Requires high-quality video and accurate scale interpretation	94.6
[21]	Utilizes facial features with LSTM and attention mechanism for MDD diagnosis	May be limited by the subtlety and variability of facial expressions in depression	91.67

3 Data Preparation

There is a basic relationship between the quality of the data that is fed into machine learning models and the integrity and robustness of the models themselves. Due to the fact that we were aware of this, we decided to commence a thorough data preparation process in order to guarantee that our dataset was suitable for the development of a trustworthy predictive model for the identification of depression.

Data Collection

During the period beginning January 1, 2015 and ending January 2, 2023, we gathered a considerable body of textual data that encompassed a wide range of self-expressed sentiments relating to personal experiences, ideas, and emotions. The purpose of the data collection was to concentrate on textual representations that are indicative of depressive states. The objective was to acquire a sophisticated understanding of the language that is utilized by persons when they are discussing depression.

Data Cleaning

Given the organic nature of user-generated content, our initial step involved cleaning the dataset to remove any irrelevant information, such as HTML tags, URLs, non- alphanumeric characters, and extraneous whitespace. This process was critical to reduce noise and improve the focus on meaningful text.

Text Normalization

After that, we normalized the text data, which entailed transforming all of the letters to lowercase. This was done to guarantee uniformity and to prevent the same words in various circumstances from being classified as separate entities.

Tokenization and StopWords Removal

The cleaned text was then tokenized, breaking down the flow of text into individual elements or 'tokens'. This step was crucial for the granular analysis of the language used. We also removed commonly used words—StopWords—that carry minimal informational weight to reduce the dimensionality of our data and highlight words more indicative of depressive language.

Lemmatization

Lemmatization was employed to consolidate different inflected forms of a word into a single item based on the word's lemma. This process is vital for reducing complexity and homogenizing variations in the text.

Text to Numeric Conversion

We then converted the text data to a numerical format, a prerequisite for feeding the data into our deep learning models. The Word2Vec method was chosen for this purpose, capturing the semantic meaning of words by mapping them into a high-dimensional space where semantically similar words are proximal to each other.

Dataset Splitting

After everything was said and done, the data that had been prepared was separated into training and test sets. This was an essential step in determining the generalizability and performance of our models. In order to educate our models to recognize patterns of depressive language, we utilized the training set. On the other hand, the test set was utilized as a fresh and objective data source for the purpose of evaluating the effectiveness of our models. As a result of this meticulous data preparation approach, we established a solid foundation for the future modeling phase, thereby laying the groundwork for the creation of a depression detection model that is both accurate and informative.

4 Data Preprocessing

In preparation for analyzing textual data, essential preprocessing steps were implemented using Python libraries such as 'tqdm' for progress tracking, 'pandas' for data manipulation, and 'text_hammer' for text cleaning operations. This setup ensures that all data handling is both efficient and reproducible, critical for maintaining the integrity of our machine learning pipeline. Additionally, the 'beautifulsoup4' library was specifically reinstalled to the version 4.12.2 to standardize the removal of HTML tags, ensuring consistency across different text inputs.

Our preprocessing workflow consisted of several systematic transformations applied to the dataset to refine and standardize the text. Initially, all text was converted to lowercase to eliminate variations caused by case sensitivity. Subsequent steps included the removal of email addresses, HTML tags, special characters, and accented characters from the text. These operations were vital for reducing noise and focusing the analysis on the substantive content of the text. Each step was methodically applied to the dataset using the progress_apply function from 'tqdm', which visually tracks processing progress, enhancing transparency and control during data manipulation.

The culmination of these preprocessing steps resulted in a cleansed and homogenized dataset, devoid of irrelevant or distracting elements and optimized for further analysis.

This meticulous preparation is crucial as it directly influences the performance and effectiveness of the subsequent machine learning models by ensuring that they are trained on data that is as relevant and informative as possible.

Model Evaluation

To ascertain the efficacy of our predictive models in accurately identifying depression from text, we conducted a thorough evaluation using standard performance metrics: Precision, Recall, and F1-Score. Each metric provides insights into different aspects of model performance, collectively offering a comprehensive overview of their capabilities.

Precision measures the accuracy of positive predictions, denoting the proportion of correctly identified depressive posts among all posts labeled as depressive by the model. Recall, or sensitivity, indicates the ability of the model to capture all relevant instances, reflecting the percentage of actual depressive posts that were correctly identified. The F1-Score harmonizes Precision and Recall into a single metric, calculating their harmonic mean to account for models that may excessively prioritize one metric over the other.

The models evaluated in this study included various deep learning architectures known for their proficiency in handling sequential and text-based data—specifically, LSTM, Bi-LSTM, and Bi-LSTM-RNN. Each model was trained using the same dataset to ensure comparability of results.

The evaluation process involved a k-fold cross-validation technique to mitigate any bias from the model's training and testing on the same data subset. This approach not only enhances the robustness of our evaluation but also ensures that our findings are generalizable across different subsets of data. For each fold, the models were trained on a designated portion of the dataset, with the remaining part used for testing. The performance metrics were then averaged across all folds to provide a final measure of model efficacy.

Our findings revealed that the Bi-LSTM-RNN model achieved the highest F1-Score among the tested models, indicating a superior balance between Precision and Recall. This model demonstrated a robust ability to discern complex language patterns associated with depression, likely benefiting from its bidirectional architecture, which processes text from both forward and backward directions, capturing a richer context. On the other hand, the standard LSTM model, while slightly lower in overall F1- Score, showed remarkable Recall, suggesting its potential utility in applications where missing a depressive post could have serious implications.

These results underscore the potential of deep learning models in mental health applications, particularly in the automated monitoring and analysis of depression-related text. Furthermore, the comparative performance across different models provides valuable insights into the strengths and limitations of each architecture, guiding future research and application development in the field of NLP-driven mental health analysis.

5 Results and Discussion

Model Performance

The core objective of our study was to evaluate the effectiveness of various deep learning models in identifying depressive content in text data. The models considered included LSTM, Bi-LSTM, and Bi-LSTM-RNN. Performance metrics such as accuracy, precision etc. were calculated to gauge each model's capability.

LSTM: Demonstrated commendable recall, making it highly effective at identifying most true positive cases. It registered an accuracy of 89.6%, precision of 90.3%, recall of 93.2%, and an F1-score of 91.7%. The Fig. 5 shows an ROC curve for an LSTM model, achieving an AUC of 0.97, which signifies excellent model performance in distinguishing between classes. The model vastly surpasses random guessing, as illustrated by the curve's proximity to the top left corner, indicative of a high true positive rate and a low false positive rate. The Fig. 6 confusion matrix indicates the highest FN (411) and the lowest TP (5603) among the three models, alongside a TN of 5386 and the highest FP (600). This indicates that while the LSTM model is still performing well, it may have a higher tendency to misclassify both depressive and non-depressive posts compared to the other models.

Bi-LSTM: Showed superior precision and very high recall, performing slightly better than the LSTM in balancing these metrics Fig. 1. The model achieved an accuracy of 91.2%, precision of 90.8%, recall of 92.9%, and an F1-score of 91.8%. The Fig. 4 ROC curve depicts the Bi-LSTM model's performance in classifying depressive content, with an AUC of 0.97 indicating a high level of accuracy. The curve's steep ascent and plateau near the top-left corner reflect a high true positive rate with a low false positive rate, outperforming random chance which is represented by the diagonal line. The Fig. 7 displays a similar pattern of strong TP (5586) and TN (5419) performance. However, it has fewer FNs (428) compared to the Bi-LSTM-RNN model but a slightly higher FP count (567). This suggests the Bi-LSTM model is slightly more sensitive in detecting depressive content, favoring fewer missed cases of depression at the expense of incorrectly labeling some non-depressive posts as depressive.

Bi-LSTM-RNN: This model combined the strengths of bidirectional processing with recurrent neural networks, leading to the highest accuracy and F1-score among the tested models. It recorded an accuracy of 92.3%, precision of 91.3%, recall of 91.8%, and an F1-score of 91.6%. The Fig. 3 is a ROC curve that illustrates the diagnostic ability of the Bi-LSTM-RNN model. With an Area AUC of 0.97, the model demonstrates excellent performance in distinguishing between depressive and non-depressive text content, significantly better than random guessing (diagonal dashed line), which has an AUC of 0.5. The Fig. 7 the confusion matrix shows a high number of true positives (TP) and true negatives (TN), 5522 and 5461 respectively, indicating accurate predictions for both depressed and non-depressed classes. However, there are 492 false negatives (FN), where the model incorrectly predicted non-depression for posts that are actually indicative of depression, and 525 false positives (FP), where non-depressive posts were misclassified as depressive. The high TP and TN suggest that the Bi-LSTM-RNN model is proficient at distinguishing relevant patterns in the text data.

Comparative Analysis

A comprehensive analysis of a number of different categorization models was carried out by us in order to determine which method would be the best suitable for our dataset. In the comparison, both conventional machine learning methods, such as Logistic Regression and Random Forest, and more complex neural network designs, such as LSTM and its variants Bi-LSTM and Bi-LSTM-RNN, were taken into consideration. Accuracy, precision, recall, and F1-score were the four metrics that were included in the evaluation criteria. The findings, which are depicted in Fig. 2, illustrate how each model performed in relation to the aforementioned measuring criteria. On the other hand, the Random Forest approach demonstrated improvements across the board, while the Logistic Regression model offered a solid foundation for benchmarking. When compared to the other neural network models, the Bi-LSTM and Bi-LSTM-RNN demonstrated competitive performance. The latter model achieved slightly higher scores in Precision and F1-score, which is suggestive of its ability to effectively balance precision and recall. In accordance with the standalone performance metrics that were described before, the LSTM model displayed high Recall and F1-scores, highlighting its usefulness in situations where reducing the number of false negatives is of the utmost importance.

The LSTM model, despite having the lowest accuracy, offers substantial benefits in scenarios where the cost of missing a positive instance (false negative) is high, due to its highest recall rate. This trait makes it particularly useful for preliminary screenings where capturing as many potential depressive instances as possible is more critical than precision.

Discussion of Implications

The findings from this study underscore the potential of using deep learning models for automatic depression detection in text data. These models can significantly assist mental health professionals by providing preliminary assessments and monitoring public mental health trends. Furthermore, the differences in model performance highlight the importance of choosing the right model based on specific needs—whether the priority is reducing false negatives (high recall) or increasing the certainty of predictions (high precision).

In conclusion, while Bi-LSTM-RNN offers the best overall performance for balanced tasks, the choice of model should be aligned with specific application goals in mental health monitoring. Future research might explore combining model predictions or integrating ensemble methods to leverage the strengths of each model type, potentially enhancing both precision and recall (Fig. 8).

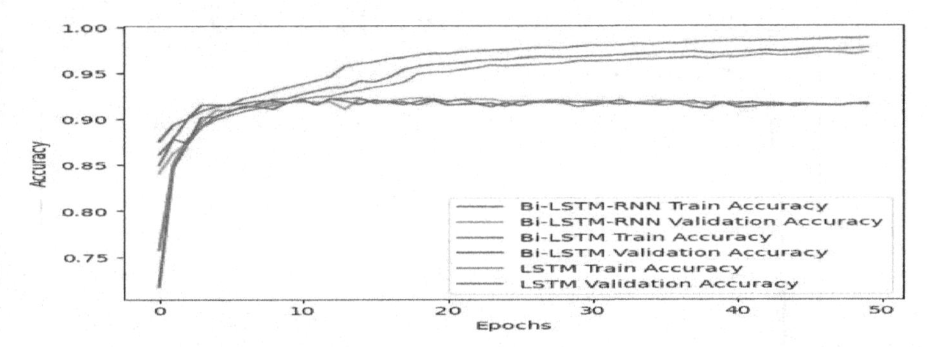

Fig. 1. Circadian Cycle Acrophase Distribution by Group

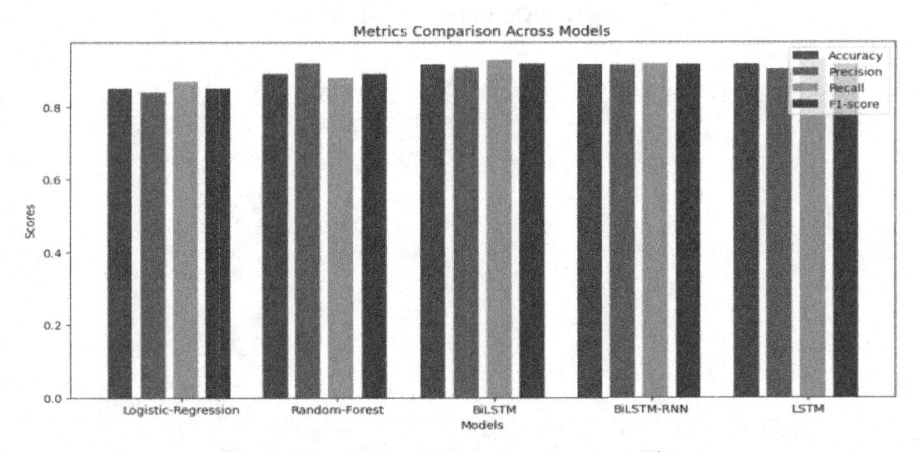

Fig. 2. Distribution of Circadian Acrophase Times

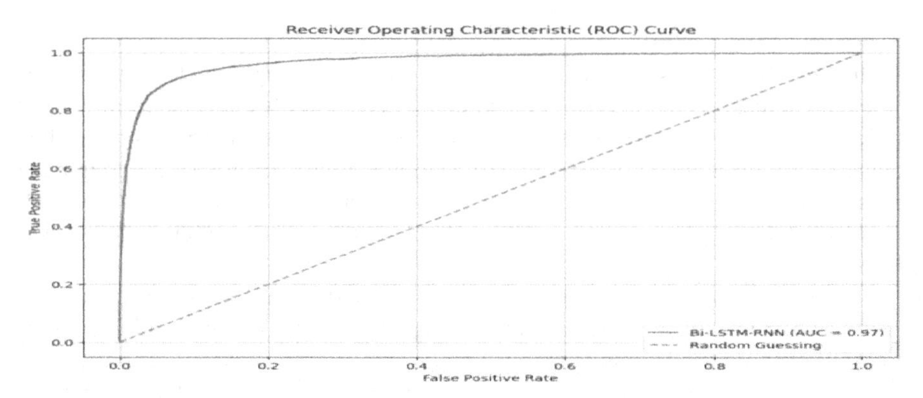

Fig. 3. ROC Curve for Bi-LSTM-RNN

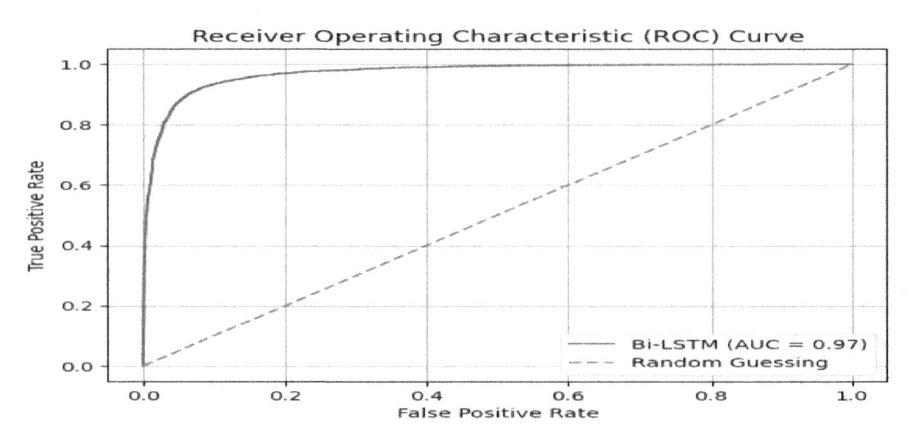

Fig. 4. ROC Curve for Bi-LSTM

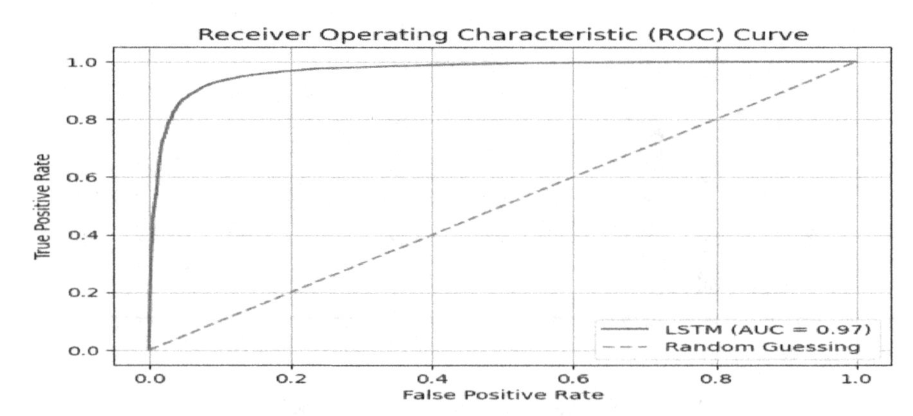

Fig. 5. ROC Curve for LSTM

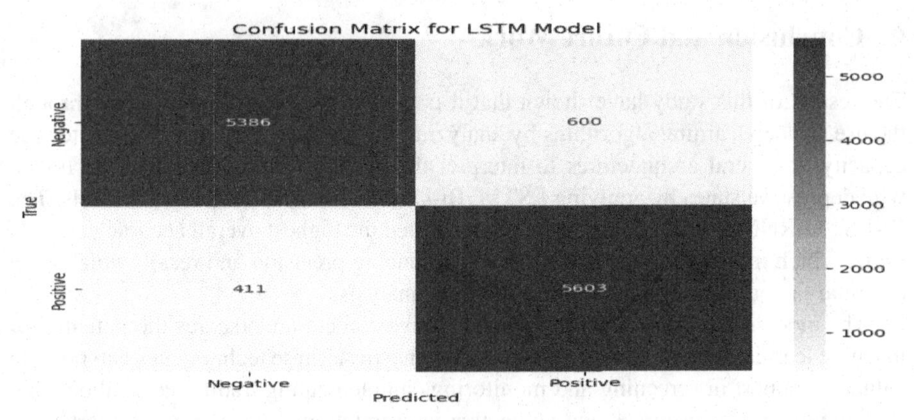

Fig. 6. Confusion Matrix for LSTM Model

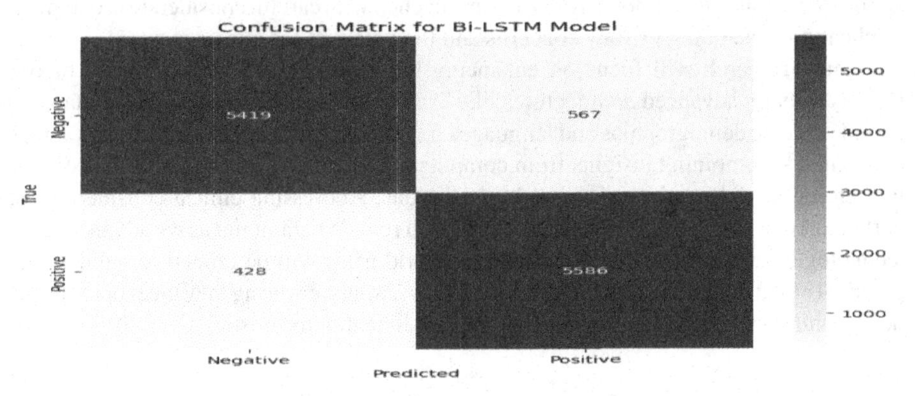

Fig. 7. Confusion Matrix for Bi-LSTM Model

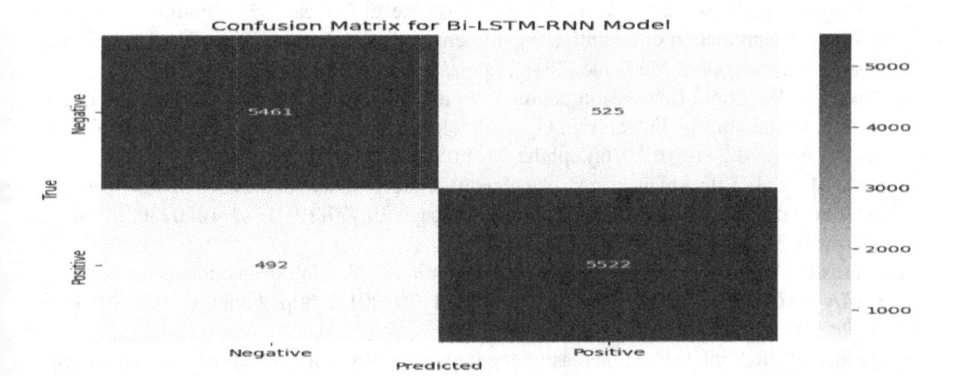

Fig. 8. Confusion Matrix for Bi-LSTM-RNN Model

6 Conclusion and Future Work

The results of this study have shown that it is possible to identify depression through the use of deep learning algorithms by analyzing textual data. We have investigated the capacity of several architectures to interpret the nuances of language that are linked with depressive states by applying LSTM, Bi-LSTM, and Bi-LSTM-RNN models. The Bi-LSTM-RNN model, in particular, demonstrated the highest overall accuracy and F1-score, which indicates that it is effective in balancing precision and recall, which is an essential factor in the context of mental health analysis.

The models' success in identifying depressive content underscores the potential of machine learning as a tool in mental health diagnostics. These technologies can provide valuable support in screening and monitoring, supplementing traditional methods that may require more resources and subjective interpretation. However, it's important to recognize that while these models perform well on the dataset provided, the real-world application of such technologies must be approached with careful consideration of ethical implications, including privacy concerns and the potential for misdiagnosis.

Future research will focus on enhancing the robustness and applicability of our models through advanced architectures like Transformers, and expanding the dataset to include diverse demographics and languages for global relevance. Collaborations across disciplines—combining insights from computer science, psychology, and linguistics—will enrich model interpretations and applications. Addressing ethical considerations, particularly around privacy and data security, will remain paramount as we advance these technologies. Additionally, conducting real-world trials will be crucial to validate the practical effectiveness of these models in clinical settings, ensuring they meet operational needs while providing tangible benefits in mental health diagnostics.

References

1. Schleiger, E.: Investigating the ability of post-stroke EEG measures of brain dysfuntion to inform early prediction of cognitive impairment or depression outcomes. The University of Queensland, School of Medicine (2016). https://doi.org/10.14264/uql.2016.613
2. Rovner, B.W., et al.: Low vision depression prevention trial in age-related macular degeneration: a randomized clinical trial. Ophthalmology (Rochester, Minn.) **121**(11), 2204–2211 (2014). https://doi.org/10.1016/j.ophtha.2014.05.002
3. Berk, M., et al.: Effect of aspirin vs placebo on the prevention of depression in older people: a randomized clinical trial. JAMA Psychiatry (Chicago, Ill.) **77**(10), 1012–1020 (2020). https://doi.org/10.1001/jamapsychiatry.2020.1214
4. Beardslee, W.R., et al.: Prevention of depression in at-risk adolescents: longer-term effects. JAMA Psychiatry (Chicago, Ill.) **70**(11), 1161–1170 (2013). https://doi.org/10.1001/jamapsychiatry.2013.295
5. Cuijpers, P., Reynolds, C.F.: Increasing the impact of prevention of depression—new opportunities. JAMA Psychiatry (Chicago, Ill.) **79**(1), 11–12 (2022). https://doi.org/10.1001/jamapsychiatry.2021.3153
6. Saba, T., Khan, A.R., Abunadi, I., Bahaj, S.A., Ali, H., Alruwaythi, M.: Arabic speech analysis for classification and prediction of mental illness due to depression using deep learning. Comput. Intell. Neurosci. **2022**, 8622022–8622029 (2022). https://doi.org/10.1155/2022/8622022

7. He, L., et al.: Deep learning for depression recognition with audiovisual cues: a review. Inf. Fusion **80**, 56–86 (2022). https://doi.org/10.1016/j.inffus.2021.10.012
8. He, L., Guo, C., Tiwari, P., Su, R., Pandey, H.M., Dang, W.: DepNet: an automated industrial intelligent system using deep learning for video-based depression analysis. Int. J. Intell. Syst. **37**(7), 3815–3835 (2022). https://doi.org/10.1002/int.22704
9. Amanat, A., et al.: Deep learning for depression detection from textual data. Electronics (Basel) **11**(5), 676 (2022). https://doi.org/10.3390/electronics11050676
10. Tian, H., Zhu, Z., Jing, X.: Deep learning for depression recognition from speech. Mob. Netw. Appl. (2023). https://doi.org/10.1007/s11036-022-02086-3
11. Cao, B., et al.: Brain morphometric features predict depression symptom phenotypes in late-life depression using a deep learning model. Front. Neurosci. **17**, 1209906 (2023). https://doi.org/10.3389/fnins.2023.1209906
12. Thakre, T.P., Kulkarni, H., Adams, K.S., Mischel, R., Hayes, R., Pandurangi, A.: Polysomnographic identification of anxiety and depression using deep learning. J. Psychiatr. Res. **150**, 54–63 (2022). https://doi.org/10.1016/j.jpsychires.2022.03.027
13. Vasconcelos, V., Domingues, I., Paredes, S.: Depression detection using deep learning and natural language processing techniques: a comparative study. In: Progress in Pattern Recognition, Image Analysis, Computer Vision, and Applications, vol. 14469, pp. 327–342. Springer (2023). https://doi.org/10.1007/978-3-031-49018-7_24
14. Bertl, M., Bignoumba, N., Ross, P., Yahia, S.B., Draheim, D.: Evaluation of deep learning-based depression detection using medical claims data. Artif. Intell. Med. **147**, 102745 (2024). https://doi.org/10.1016/j.artmed.2023.102745
15. Sami Khafaga, D., Auvdaiappan, M., Deepa, K., Abouhawwash, M., Khalid Karim, F.: Deep learning for depression detection using twitter data. Intell. Autom. Soft Comput. **36**(2), 1301–1313 (2023). https://doi.org/10.32604/iasc.2023.03336
16. Almars, A.M.: Attention-based Bi-LSTM model for Arabic depression classification. Comput. Mater. Continua **71**(2), 3091 (2022). https://doi.org/10.32604/cmc.2022.022609
17. Singh, J., Wazid, M., Singh, D.P., Pundir, S.: An embedded LSTM based scheme for depression detection and analysis. Procedia Comput. Sci. **215**, 166–175 (2022). https://doi.org/10.1016/j.procs.2022.12.019
18. Zhang, Y., Fu, Z.: The study of EEG recognition of depression on Bi-LSTM based on ERP P300. E3S Web Conf. **185**, 2007 (2020). https://doi.org/10.1051/e3sconf/20201850200
19. Zhao, Y., Xie, Y., Liang, R., Zhang, L., Zhao, L., Liu, C.: Detecting depression from speech through an attentive LSTM Network. IEICE Trans. Inf. Syst. **E104.D**(11), 2019–2023 (2021). https://doi.org/10.1587/transinf.2020EDL8132
20. Xie, W., et al.: Multimodal fusion diagnosis of depression and anxiety based on CNN-LSTM model. Comput. Med. Imaging Graph. **102**, 102128 (2022). https://doi.org/10.1016/j.compmedimag.2022.102128
21. Mahayossanunt, Y., Nupairoj, N., Hemrungrojn, S., Vateekul, P.: Explainable depression detection based on facial expression using LSTM on attentional intermediate feature fusion with label Smoothing. Sensors (Basel, Switzerland) **23**(23), 9402 (2023). https://doi.org/10.3390/s23239402
22. Pradhan, N.R., Ghosh, H., Rahat, I.S., Naga Ramesh, J.V., Yesubabu, M.: Enhancing Agricultural Sustainability with Deep Learning: A Case Study of Cauliflower Disease Classification. https://doi.org/10.4108/eetiot.4834
23. Ghosh, H., Tusher, M.A., Rahat, I.S., Khasim, S., Mohanty, S.N.: Water quality assessment through predictive machine learning. In: Intelligent Computing and Networking. IC-ICN 2023. Lecture Notes in Networks and Systems, vol 699. Springer, Singapore (2023). https://doi.org/10.1007/978-981-99-3177-4_6

24. Rahat, I.S., Ghosh, H., Shaik, K., Khasim, S., Rajaram, G.: Unraveling the heterogeneity of lower-grade gliomas: deep learning-assisted flair segmentation and genomic analysis of brain MR images. EAI Endorsed Trans. Perv. Health Tech. [Internet] **9**, (2023). https://doi.org/10.4108/eetpht.9.4016
25. Ghosh, H., Rahat, I.S., Shaik, K., Khasim, S., Yesubabu, M.: Potato leaf disease recognition and prediction using convolutional neural networks. EAI Endorsed Scal. Inf. Syst. [Internet] **21**, (2023). https://doi.org/10.4108/eetsis.3937
26. Mandava, M., Vinta, S.R., Ghosh, H., Rahat, I.S.: An all-inclusive machine learning and deep learning method for forecasting cardiovascular disease in Bangladeshi population. EAI Endorsed Trans. Perv. Health Tech. **9**, (2023)
27. Mandava, M., Vinta, S.R., Ghosh, H., Rahat, I.S.: Identification and categorization of yellow rust infection in wheat through deep learning techniques. EAI Endorsed Trans. IoT **10**, (2023). https://doi.org/10.4108/eetiot.4603
28. Khasim, S., Rahat, I.S., Ghosh, H., Shaik, K., Panda, S.K.: Using deep learning and machine learning: real-time discernment and diagnostics of rice-leaf diseases in Bangladesh. EAI Endorsed Trans. IoT **10**, (2023). https://doi.org/10.4108/eetiot.4579
29. Khasim, S., Ghosh, H., Rahat, I.S., Shaik, K., Yesubabu, M.: Deciphering microorganisms through intelligent image recognition: machine learning and deep learning approaches, challenges, and advancements. EAI Endorsed Trans. IoT **10**, (2023). https://doi.org/10.4108/eetiot.4484
30. Mohanty, S.N., Ghosh, H., Rahat, I.S., Reddy, C.V.R.: Advanced deep learning models for corn leaf disease classification: a field study in Bangladesh. Eng. Proc. **59**, 69 (2023). https://doi.org/10.3390/engproc2023059069

Cognitive Computing in Cyber Physical Systems: A Robust Computational Strategy for Anomaly Detection

K. S. Aakaash, N. D. Patel, and Ajeet Singh(✉)

School of Computing Science and Engineering (SCSE),VIT Bhopal University, Sehore 466 114, Madhya Pradesh, India
{aakaash.k2020,narottamdaspatel,ajeetsingh}@vitbhopal.ac.in

Abstract. By monitoring system activity and categorising it as either normal or anomalous, an anomaly-based intrusion detection system can identify computer and network intrusions as well as misuse. In Cyber Physical Systems, anomaly-based intrusion detection is a crucial task as it can help prevent system failures, ensure safety, and optimise performance. This research work evaluates the effectiveness of various Machine Learning models for anomaly-based intrusion detection in training using the confusion matrix of expected outcomes. Elliptic Envelope, Isolation Forest, and One-class SVM anomaly detection techniques are utilized here. Subsequently, the models underwent training, testing on input data, and parameter tuning to achieve the peak performance. The results with the functionality of the Elliptic Envelope seems to be lacking, especially in approach 2. In approach 1, the combination of the OC-SVM and Isolation Forest takes into consideration of individual variables. On the other hand, approach 2 suggests that the OC-SVM should be prioritized due to the intricate nature of matching algorithms. This analysis aims to enhance the capacity to identify intrusions and protect continuous processes from deviations.

Keywords: Cyber Physical Systems · One-Class SVM · Anomaly-based Intrusion Detection · Isolation Forest · Dimensionality Reduction · Elliptic Envelope · Principal Component Analysis

1 Introduction

A class of complex systems known as Cyber Physical Systems (CPS) combines physical, networking, and computational activities. They are made up of two basic parts: the physical parts, which are the system's physical processes and components, and the Cyber parts, which are the system's computing processes and components. In CPS, anomaly detection is a crucial instrument for protecting against cyberattacks. Anomaly detection in CPS is deliberately positioned amongst the changing landscape of the modern technology world. Through its ability to weave computational algorithms and physical processes CPS has

S. Tiwari et al. (Eds.): AI4S 2024, CCIS 2243, pp. 225–245, 2025.
https://doi.org/10.1007/978-3-031-81369-6_18

become vital to run a multiplicity of applications in diverse fields including smart cities and healthcare to industrial automation and transportation [1]. Therefore, the necessity for ensuring that such systems are secure from mischief increases. The complexity of CPS is drawn from their inter-connectivity and their dynamic nature, which exceed the capabilities of conventional anomaly detection algorithms. Traditionally, rule-based processes and statistical methods were applied; however, the emergence of CPS mandates a new approach to the matter. This paradigm shift which was primarily designed to unlock the potential of cognitive computers that seek to merge machine learning and artificial intelligence finds the impetus in the domain of anomaly detection which now combines adaptability, intelligence, and foresight [2].

This paper sets out a broad review, which explores the aforementioned topic from various aspects, thus highlighting the issues, difficulties and achievements in this research area. Through the navigation of the web of technologies, theoretical frameworks, and practical considerations, we unravel the role of cognitive computing as a catalyst for the rewriting of the resultant anomaly detection strategy. Beyond its immediate scope of cybersecurity, this study aims to elucidate the wider issues, such as resilience, efficiency and the sustainability of our cyber-physical ecosystems, which underpin our increasingly interdependent society. Attacks on CPS have increased as a result of its quick growth as mentioned in [3]. Finding anomalies is crucial to identifying and stopping these attacks. Finding odd patterns or occurrences in data that may point to malicious behavior is known as anomaly detection. Malicious activity like as denial of service assaults, malicious code execution, and reconnaissance can be identified by anomaly detection. Because CPS are more interconnected and dependent on one another, they are more susceptible to hackers. The complexity of the system grows with the number of connected devices, making it more challenging to quickly identify and stop intrusions. This issue can be effectively resolved with anomaly detection, which identifies unusual network activity and notifies administrators of any dangers. Large-scale data is needed for cybersecurity analysis in order to forecast, recognize, describe, and address risks.

Systems for anomaly detection have a lot to offer CPS security. By detecting and responding to hostile acts in real-time, these technologies can help reduce the risk of errors and help recover from and minimize the harm caused by cyber-attacks [3]. Yet another concern is the difficulty in scaling the approach to the detection of anomalies, given their vastness and intricacy in connected cyber-physical systems (CPS). The vast amount of data which is generated by sensors, devices, and interconnected systems creates a huge challenge in the form of compute power, capacity and efficiency. Cognitive computing being so complex, is faced with the scalability issue though it offers great functionality. This should be addressed before the integration of cognitive computing into large-scale CPS applications. Effective algorithms, distributed computing frameworks, and hardware infrastructure advanced are the critical domains in addressing this challenge which is necessary for the cognitive computing to be applied in anomaly detection at a large scale across different CPS applications.

The urgency of cyber protection is too great for human efforts to handle as data volume and complexity rise. Advances in computer data storage, handling, and acquisition have made it possible for artificial intelligence (AI) and machine learning (ML) to identify complicated patterns and trends more quickly and effectively than humans. Despite ongoing advancements in the security infrastructure of Industrial Control Systems (ICS), there exists a potential for the convergence of Machine Learning and Data Science applications, which could enhance our ability to detect process deviations [4]. CPS are getting more and more necessary in the current era. Smart grids, driverless cars, and industrial control systems like water plants are a few examples of CPS. Because CPS depend on reliable and consistent data to operate, they are highly susceptible to abnormalities and errors that might corrupt this data and compromise the system's overall dependability. Systems for anomaly detection have a lot to offer CPS security. By detecting and responding to hostile acts in real-time, these technologies can help reduce the risk of errors and help recover from and minimize the harm caused by cyberattacks. Additionally, network visibility can be provided via anomaly detection systems, informing administrators of potential dangers and weak points. With this network awareness, administrators could be better equipped to spot malicious activity and take the appropriate action.

1.1 Research Challenges

The core research challenges are discussed in this section.

High Dimensionality: The enormous dimensionality of the data is one of the main obstacles to anomaly detection in cyber-physical systems. Data that has a lot of features or variables is referred to as high-dimensional data, which can make it challenging to effectively identify the underlying trends and discern between normal and abnormal behavior. This is due to the fact that when the number of dimensions in the data increases exponentially with the number of features, it becomes increasingly difficult to identify patterns and connections between the variables. Many factors can contribute to the enormous dimensionality of data in cyber-physical systems.

One of the reasons is the vast array of sensors and other data sources that are regularly used to keep an eye on these systems. A power grid, for instance, might contain hundreds or even thousands of sensors dispersed across the system, each of which produces a time series of measurements. Due to the fact that each sensor produces a unique data series, the high number of sensors results in a high-dimensional data space. The duration of the time series is another element that may contribute to the high dimensionality of data in cyber-physical systems. These systems frequently have very long time series data with hundreds or even millions of data points. As a result, the data may become more dimensional as each data point may represent a distinct feature. Cyber physical systems' large data dimensionality might make it challenging to effectively model the data and spot unusual behavior. One-class SVMs and kernel density estimation are two common traditional approaches for anomaly identification that may struggle to handle high-dimensional data and may not be reliable enough to effectively capture the complex patterns and relationships in the data.

Lack of Labelled Data: The majority of the time, detecting anomalies involves unsupervised learning, which means that the training data does not contain labels indicating whether a specific sample is normal or abnormal. The model must therefore learn to differentiate between typical and abnormal behavior based on the data itself, which might be difficult if the data is very volatile or comprises a sparse collection of anomalous examples. There are several reasons why labeled data is missing from cyber-physical systems. One factor is the systems' complexity, which could make it difficult to correctly identify the data. It can be difficult to determine if a given data instance is normal or abnormal since a number of factors, including weather and equipment breakdowns, can affect how a power system behaves. The cost and duration involved in labeling the data is another aspect. Data labeling frequently calls for the knowledge of subject matter experts, who could be hard to come by or perhaps too busy to classify a massive amount of data. Additionally, categorizing the data can take a while, particularly for systems with a lot of data series or a lengthy time horizon.

1.2 Contribution Highlights of the Paper

- The cognitive computing-based similarity connects to the exploration of anomaly detection strategies applied to cyber-physical systems, thus exponentially faster then the divided network.
- Developed a new perspective on how to detect anomalies. Reversed the trend of researchers going back to older and current intelligence techniques, which in turn influenced the entire field of research.
- Bypassed the obstacle of scalability in the domain of fault detection via development of a reliable set of algorithms that are capable of handling the complexity and massiveness of data that is produced by the modern cyber-physical systems.
- Conducted experimental evaluation to make sure that the development and the implementation of proposed strategies are leading to change in the paradigm of anomaly detection, as a strong empirical base supports a paradigm break.

1.3 Structure of the Paper

Beginning with Sect. 1, this exploration is focused on Machine Learning's ability to give real time process status updates within industries of process. It also focuses at the issues faced during implementation which are necessary for overcoming obstacles and ensuring smooth integration. This has been followed by Sect. 2, which carries out a literature survey taking from past processes and core developments in industrial and research perspectives and provides a base for understanding certain research gaps. Further, it leads us into Sect. 3 in which we discuss about adapted methodologies and experimental evaluation. Section 4 presents the obtained simulation results, benchmarking and discussions. Finally, Sect. 5 provides conclusive discussion and future research scope.

2 Literature Survey

In preparation for this literature review on anomaly detection in cyber physical systems, it is imperative to acknowledge the existence of several comprehensive surveys and reviews in this field. This section presents a summary of the most relevant and recent sources available. These resources will serve as a foundation for comprehending the advancements made in the field and provide insight into the challenges and opportunities for future research. A number of papers discuss supervised learning algorithms, such as Naive Bayes, k-Nearest neighbour classifiers, Support Vector Machines (SVM), Multi Layer Perceptron (MLP), Decision Trees, k-Nearest neighbour classifiers, Neural Network, Random Forest, Random Tree [5]. Zhang et al. proposed a vital instrument for guaranteeing the dependable and effective operation of contemporary power grids is anomaly detection. Anomaly detection techniques can be used to recognize and stop anomalous phenomena, such as odd consumer consumption patterns, malfunctioning grid infrastructure, outages, outsider cyberattacks, or energy fraud [6]. Work [7], authored by Pang et al. offers a thorough overview of the research being done in the field of deep anomaly detection. In many academic communities, anomaly detection-also referred to as outlier or novelty detection-has long been a study topic. The authors' goal is to present a thorough summary of the most recent developments in this field, with an emphasis on deep learning-based strategies.

A study by [8] explored the exquisite susceptibility of Decision Trees to training data, especially when faced with imbalanced datasets. They also noticed that not only Decision Trees, but even Random Forests which is in fact an ensemble of decision trees is greatly affected by the training data on which it has been built. Its performance can vary widely from one set to another. This contrasts with the Naive Bayes Classifier, which uses Bayes theorem to make highly efficient classifications. Its training process is also less time-consuming than that of MLP classifiers as mentioned in [9].

Taormina et al. introduces a semi-supervised approach for detecting and localising cyberattacks in water distribution systems. This approach utilises maximum canonical correlation analysis (MCCA) to reduce the dimensionality of the problem and support vector data description (SVDD) to classify anomalous samples without needing labeled attacks in the training dataset. The technique showed consistently good performance in identifying and localizing cyberattacks when evaluated on two case studies and multiple datasets. Neglecting to identify irregularities in a CPS could have detrimental effects. For instance, undiscovered anomalies may cause system failures, which may seriously harm the system and its parts. Undiscovered abnormalities can also endanger human lives in safety-critical systems, such as those utilised in transportation, healthcare, energy or critical infrastructure such as water treatment facilities [10]. Wang et al. proposed a enhanced Support Vector Machine (SVM) performance by incorporating Principal Component Analysis (PCA) and Particle Swarm Optimization (PSO) [11]. This strategy, which uses PCA to reduce data dimensions and employs PSO for optimizing kernel parameters, performed significantly better than standard SVM as well as its counterparts using PSO. This paper presents an innovative approach that combines different machine learning and data mining techniques to produce numerous invariant rules for ICS anomaly detection, as proposed by [12].

Karnouskos & Stamatis examined the extremely complex features of Stuxnet and how it affects current security considerations to provide ideas for the next generation of SCADA (Supervisory Control and Data Acquisition) systems from a security standpoint [13]. In order to identify abnormalities and stop attacks, it emphasizes how crucial it is to integrate risk analysis and security technologies into industrial systems. Additionally, it suggests that security should be integrated into all operational aspects of industrial systems and that industry should invest in upgrading the security of their systems to ensure the safety of their processes and critical infrastructure. Another a more relevant example can be seen in [14]. In this paper, a graphical model-based approach is proposed for detecting anomalies in the operation of an Industrial Control System (ICS) called SWaT (Secure Water Treatment) similar to [15]. The approach involves learning timed automata as a model of normal behavior based on sensor signals, and learning Bayesian networks to discover dependencies between sensors and actuators. The learned models are then used as a one-class classifier for anomaly detection, allowing the detection of irregular behavioral patterns and dependencies. The approach is applied to a dataset collected from SWaT and is shown to outperform other methods such as support vector machines and deep neural networks in terms of precision and run-time. The approach is also interpretable, allowing for the localization of abnormal sensors or actuators.

3 Methodology Pipeline and Experimental Evaluation

The methodology's pipeline in Fig. 1 illustrates a proposed framework for anomaly detection in Cyber-Physical Systems (CPS). The process begins with the collection of attack data from the WADI dataset, which consists of various sensor readings (e.g., 1_AIT_001_PV, 1_FIT_001_PV, etc.). In the Pre-Processing Stage, null names are removed, followed by normalization and Principal Component Analysis (PCA) for dimensionality reduction. The dataset is then split into training and testing sets. Two distinct approaches are used for model building: Approach 1 and Approach 2, where different models such as One-Class SVM, Elliptic Envelope, and Isolation Forest are trained. These models are subsequently evaluated based on precision, recall, F1 values, and accuracy to determine the best-performing model for anomaly detection in the CPS environment.

3.1 Dataset Description

This research work mainly focused on a dataset obtained from the Water Distribution (WADI) Plant, which was made available by the itrust Centre at the Singapore Institute of Technology and Design (SUTD). The entire procedure is divided into two sub-processes namely Approach 1 and Approach 2, also the entire dataset has been partitioned into two segments. The first segment includes the initial 12 days of standard plant operation and the second segment includes the final 2 days during which all planned 15 types of attacks were launched, resulting in approximately 10,000 data points. The aim is to maximize the attack detection with minimum false alarms raised i.e. increasing both precision and recall simultaneously.

3.2 Mechanism of the Model

The Proposed Model includes unsupervised outlier detection methods in Scikit Learn (Machine Learning Library in Python) such as One-Class SVM, Elliptic Envelope and Isolation Forest. These methods consist of some parameters that need to be altered and tuned for the most optimum performance. Training an entire dataset at once which has about 130 dimensions is time consuming and less accurate. This issue could be addressed by dividing the dataset into its subsequent processes and further analyzing them individually. Hence, Approach 1 and Approach 2 have been sliced from complete normal data. Even after slicing, the number of dimensions in both Approach 1 and 2 were still significant. For that, Principal Component Analysis (PCA) technique is used for dimensionality reduction of data. The analysis is divided into 2 parts as discussed in next subsections.

Pre-processing Stage: A strict quality control approach was used in pre-processing stage of the WADI_attackdata dataset which allowed for raw data to be curated, cleaned and then ready to be analyzed. The first phase involved the identification and deletion of places where null values were stored in the attributes of these particular variables such as 1_AIT_001_PV, 1_AIT_003_PV, 1_AIT_004_PV, and 1_FIT_001_PV. Null values were omitted to ensure dataset is being whole and unbroken. Next, though in order to establish standard form in attribution data and also to enable accurate comparisons and normalization process was carried out. By doing so, the numbers representing the chosen parameters to some standardised scale such as from 0 to 1 covering the minimum to maximum numbers were changed. Scale normalization becomes an integral stage during data pre-processing due to its impact on scaling attributes and avoiding distortion in the analysis as a result of this attributes.

Data Imbalance Handling: The absence of labeled dependent variable in the dataset presents a specific problem that makes unsupervised learning applicable.

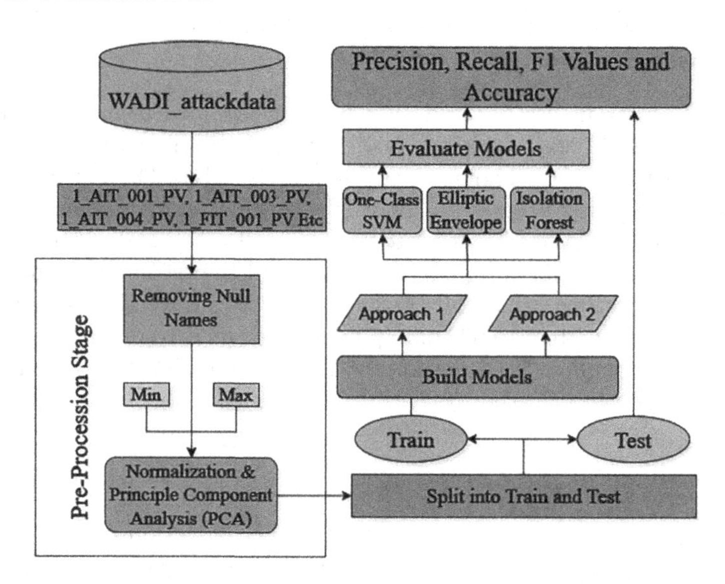

Fig. 1. Proposed Framework for Anomaly Detection in CPS

Being on top of this is even more complicated here because of the unbalanced data validation, which in turn brings the unrevealed depth of the anomaly detection task. The primary challenge of finding attacks in this context has led to a novel approach, it is the rephrasing of the analysis problem as an outlier/novelty detection problem. In this framework, only the training data set consists of normal data, and it is the job of the model to filter the validation set into inliers (normal) or outliers (abnormal).

To address this issue of unsupervised learning and for the model to accurately discriminate abnormalities, the analysis uses PCA as an important univariate technique which reduces the number of dimensions. PCA is a tool that is very important in the process of transforming data which was numerically complex to a reduced set of main components. Here alongside variance data reduction is equally essential for the building a model that can not only reduce the effect of the imbalanced data but also highly proficient in identifying the abnormalities within the unsupervised learning ecosystem. The application of PCA according to this purpose relates to the overall objective of improving the detecting anomaly accuracy and the efficiency of which are not limited in the range of labeled variables, being a vital feature of the analysis. Principles of dimensionality reduction imply that data set is simplified when it is transformed into lower dimensions, keeping only the most significant information without the less relevant or duplicate features. Therefore, this process makes the machine learning models get rid of the difficulties of the "curse of dimensionality" that could arise if there was a huge number of input elements.

Model Building: After preprocessing the WADI_attackdata dataset, it was systematically divided into training and testing sets to facilitate both the development and evaluation of a strong anomaly detection model. Three different algorithms, One-Class Support Vector Machine (One-Class SVM), Elliptic Envelope, and Isolation Forest were used to develop the model.

One class SVM algorithm is particularly competent in outing deviations due to the process of training which enables it to note characteristics of the normal data points. Hence one class SVM can be used appropriately in conditions where average readings are far more than outliers. While the Elliptic Envelope algorithm relies on the assumption that normal data points are generated from Gaussian distribution and thus determines outliers by measuring deviations from this distribution, the One-Class SVM algorithm is based on the idea that outliers are characterized by higher feature values than normal data points. In addition, the Isolation Forest algorithm is good at separating outliers since the algorithm is realized by traversing through the permutations of each tree and counting the average path length. After the model had been trained on the data it was given, the process of evaluation commenced to asses the model's performance effectively. This was done by evaluating important metrics such as precision, recall and F1-score as well as those used for clarifying ability of the model to detect anomalies while ensuring minimal false positive. The assessment process frequently brought attention to the strong or weak points of the model for data anomaly detection within WADI_attackdata set. The methodology used in this study is different, and has to adapt the Anomaly detection techniques for Self-supervised algorithms because there's no labeled target variable in WADI. The advantages of these methods are outlined as follows:

- Can be trained in a reasonable amount of time.
- Random samples can be used instead of complete dataset as the methods compute a hypothetical surface boundary for the training data.
- Sensitive to training set (i.e., customizable for the definition of inliers).

3.3 Broad Spectrum of Approaches

Approach 1 (Core Distribution and Evaluation): Approach 1 (A1) encompasses a set of 19 variables, encompassing various equipment and meters. These include transfer pumps, level indicators, motorized valves, as well as indicators for physical properties such as pH, turbidity, conductivity, TRC (Total Residual Chlorine) and ORP (Oxidation-Reduction Potential), and. With the help of an in-built module of PCA in Scikit Learn, a PCA function is defined with 95% explained variance. It means the number of principal components obtained after fitting, will altogether explain at least 95% variance of complete Approach 1 variables. After fitting and transforming PCA function on A1's normal data, it was found that only 3 PCs explained around 99.8% variance. The obtained PCA variance ratio is depicted as Fig. 2.

```
pca.explained_variance_ratio_

array([0.60433615, 0.27824022, 0.11554166])
```

Fig. 2. PCA Variance Ratio

Similarly, attack data is also transformed using the same PCA function and both normal and attack data of Approach 1, after principal component transformation, were stored in data frames. The first two principal components of both the data frames were plotted to visualize the similarity and difference in the data points (depicted as Fig. 3). This figure was plotted after manual removal of some extreme outliers, for better scalability and visuals of plot. It is observed that a large proportion of attack data log is behaving as normal, which indeed, is as expected. The dataset is now prepared to train the models.

1. **One Class SVM (OC-SVM):**
   The Scikit Learn module of OC-SVM has some parameters associated with it. One of them is 'nu' which represents the fraction of training errors. Different experiments were conducted for different values of nu. In general, the trend found is, with decreasing the fraction of training errors, more data points in attack data log has been identified as normal. Most appropriate case (nu = 0.0001) is chosen for further evaluation which identified 1,53,976 instances out of 1,72,795 as normal. The representation is provided as Fig. 4.

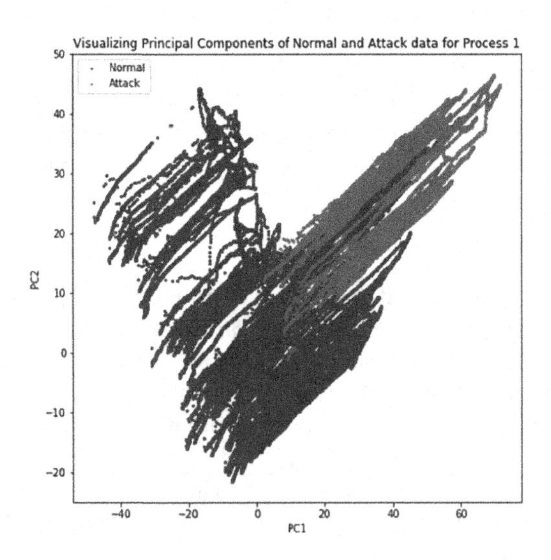

Fig. 3. Approach 1 - Visualizing Principal components

```
# Let's predict how many data points has been identified as normal in attack data.
print(np.count_nonzero(OC_SVM.predict(attack_PC)==1),'data points has been identified as normal')

153976 data points has been identified as normal
```

Fig. 4. OC-SVM Normal points

Furthermore, the points that have been detected as attacks, along with their 'date time' stamps, were plotted to understand the predictions of the model. The depiction for the same is given as Fig. 5. A pattern of inconsistency in attack detection, false alarms, delayed detection can be observed in above plotted predictions of OC-SVM model. There are some characteristics of OC-SVM model that are highlighted below:

i) **Strictly specific to the range of training data variables:** Normal data for Approach 1 had a discrepancy for a variable '1_LT_001_PV' which is the level indicator of Primary Tank 1 (Raw Water Tank 1). The readings of this indicator fluctuated very frequently, and even at many instances, found close to 100. Thus, to avoid redundancy in the model, training data were selected in such a way that the '1_LT_001_PV' variable had the range of 40–70.
ii) **High sensitivity:** This argument can be supported by the predictions of the first attack. According to the Table 1, the details of first attack is as follows:

Table 1. Details of first attack

Starting Time	Ending Time	Duration (minutes)
9/10/17 19:25:00	9/10/17 19:50:16	25.16

Above description states that the attack started at 19:25:00 but the OC-SVM model started predicting the readings as anomaly close to 19:48:00, which is a lag time of more than 20 min. This delay occurred because the '1_LT_001_PV' indicator's reading, in attack log, crossed the value of 70 at 19:47:26 as shown in the Fig. 6.
Hence, it justifies the sensitivity of OC-SVM to the training set variable's range and its impacts on detection of abnormalities in the process.

2. **Isolation Forest:** To "isolate" observations, the Isolation Forest first chooses a feature at random, and then it chooses a split value at random between the feature's maximum and minimum values. This method has two main parameters that need to be defined before training:
'contamination' - The proportion of outliers in the data set.
'n_estimators' - The number of base estimators in the ensemble.
In general, it is found that decreasing contamination increases the number of instances identified as normal (inlier). Though, increasing base estimators

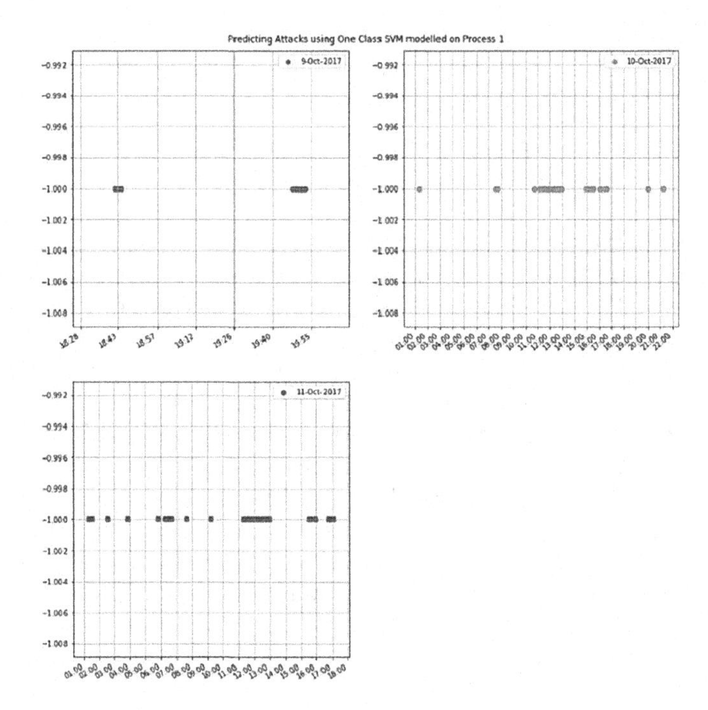

Fig. 5. OC-SVM predictions for Process 1

	Date	Time	1_LT_001_PV
6440	10/9/2017	7:47:20.000 PM	70.000
6441	10/9/2017	7:47:21.000 PM	70.000
6442	10/9/2017	7:47:22.000 PM	70.000
6443	10/9/2017	7:47:23.000 PM	70.000
6444	10/9/2017	7:47:24.000 PM	70.000
6445	10/9/2017	7:47:25.000 PM	70.000
6446	10/9/2017	7:47:26.000 PM	70.238
6447	10/9/2017	7:47:27.000 PM	70.238
6448	10/9/2017	7:47:28.000 PM	70.238
6449	10/9/2017	7:47:29.000 PM	70.238

Fig. 6. Locating the time when level indicator above 70

```
# Let's predict how many data points has been identified as normal in attack data.
print(np.count_nonzero(Isolation_Forest.predict(attack_PC)==1),'data points has been identified as normal')

162579 data points has been identified as normal
```

Fig. 7. Isolation Forest Normal Points

increases the number of inlier predictions, it also makes the algorithm computationally heavy which ultimately raises the training time. Therefore, based on the problem, an optimum balance between the parameters should be established. After conducting several experiments, an appropriate combination of *n_estimators* & *contamination* is found at 200 & 0.0001 respectively. This model identified 1,62,579 entries as normal out of 1,72,795. This is depicted as Fig. 7. The corresponding 'predicted attacks' vs 'date-time' plot is shown as Fig. 8.

Some observed facts from above plot are:

A. Isolation Forest predicted more instances as normal in comparison to the OC-SVM model (i.e., less number of false alarms).
B. Number of attacks predicted out of total attacks occured (Recall) is more in case of OC-SVM.
C. It appears that Isolation Forest is sensitive to only major deviations in the process variable's readings.

3. **Elliptic Envelope:**
   This method is for detecting outliers in a Gaussian distributed dataset. This technique is quite harsh and less effective for the current problem. It is recommended that this method might not be effective on large dimensional datasets. Similar to Isolation Forest, this module too, has a contamination parameter that needs to be defined before fitting the model. The most suitable model was trained which predicted 1,57,496 data points as normal out of 172795. It is depicted as Fig. 9. The corresponding predicted attacks versus 'date-time' plot is shown as Fig. 10.

Insights, that can be obtained from above plots are as follows:

A. Elliptic Envelope and Isolation Forest performed almost equal in detecting the data points of 9-Oct-2017.
B. Elliptic Envelope performance is least efficient, both in terms of precision and recall, as compared to the other two processes.

Approach 2 (Domestic Water Circuit with Pump Augmentation): The overall approach of analyzing Approach 2 will be very much similar to what we carried out in Approach 1. This process is the largest among these three and initially consists of about 86 variables in the raw format. However, after

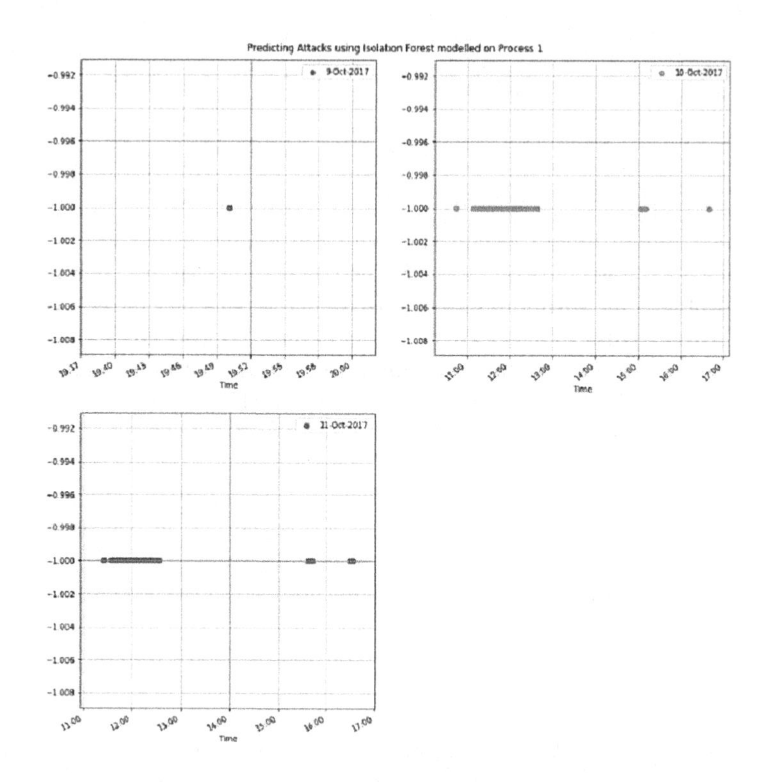

Fig. 8. Isolation Forest predictions for Process

```
# Let's predict how many data points has been identified as normal in attack data.
print(np.count_nonzero(Elliptic_Envelope.predict(attack_PC)==1),'data points has been identified as normal')

157496 data points has been identified as normal
```

Fig. 9. Elliptic Envelope Normal Points

discarding some less effective variables, we are left with a total of 75 variables that need to be passed through the PCA function. Here again, PCA is fitted for explaining 95% of the variance which resulted in 7 Principal Components. Infact, PCA reduced normal and attack dataset, at first, doesn't represent much difference as shown in plot (depicted as Fig. 11).

Training model on the above dataset is ineffective against attack detection. Hence, outliers were removed from the normal dataset while retaining 90–95% of the information. After filtering outliers from PCA reduced normal data, the plot obtained is as follows (depicted as Fig. 12).

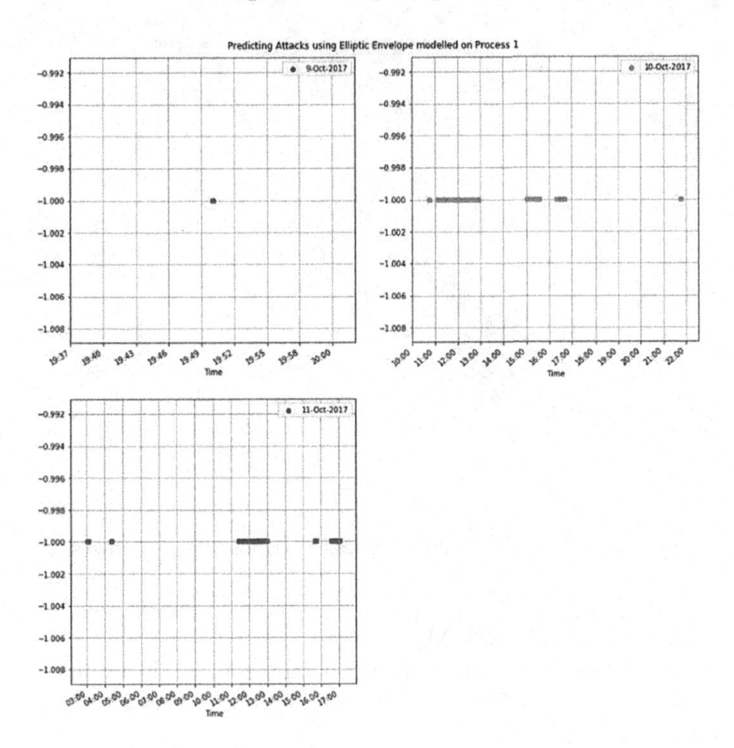

Fig. 10. Elliptic Envelope prediction for Process

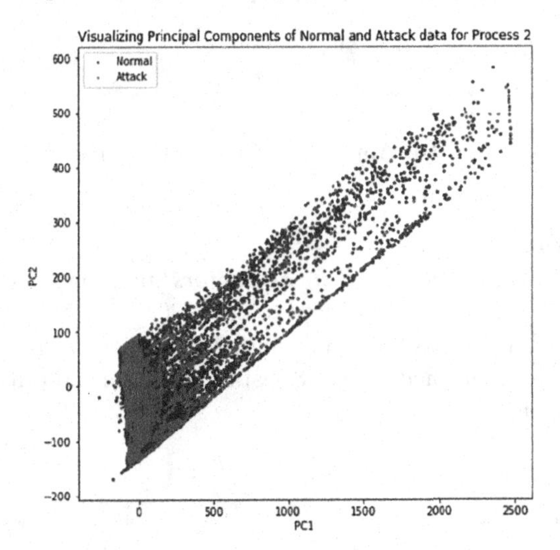

Fig. 11. Visualizing Principal Components for Process 2

Now, the normal data is eligible to be trained in different models. Beyond this point, the training and evaluation of the algorithms is conducted in the manner similar to Approach 1.

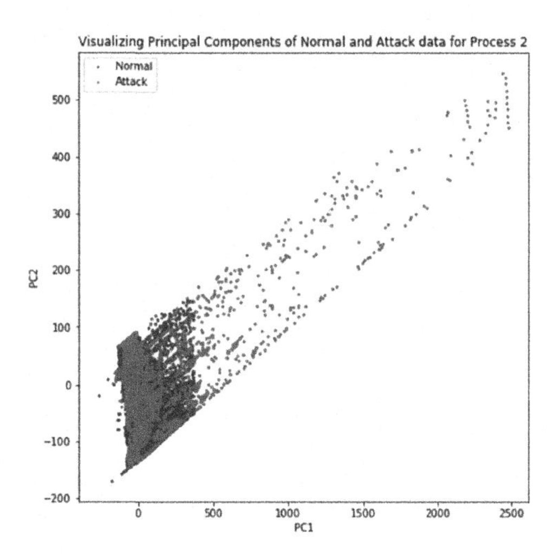

Fig. 12. Outlier Removed Principal Components for Process 2

1. **One Class SVM (OC-SVM):**
   After several experiments, most optimum value for *'nu'* is obtained as 0.0001. The model predicted 1,58,073 instances as normal out of 172801. This is depicted as Fig. 13.

```
# Let's predict how many data points has been identified as normal in attack data.
print(np.count_nonzero(OC_SVM.predict(P2_attack_PC)==1),'data points has been identified as normal')

158073 data points has been identified as normal
```

Fig. 13. Approach 2 OC-SVM Normal

2. **Isolation Forest:**
   Tuning the *'contamination'* and *'n_ estimators'* parameters through repeated trials, optimum combination is found to be 0.0001 and 150 respectively (Infact *n_ estimators* didn't have a significant impact once it crossed the value of 100). This model identified 1,68,578 instances as normal out of 172801. This is depicted as Fig. 14.

```
# Let's predict how many data points has been identified as normal in attack data.
print(np.count_nonzero(Isolation_Forest.predict(P2_attack_PC)==1),'data points has been identified as normal')

168578 data points has been identified as normal
```

Fig. 14. Approach 2 Isolation Forest Normal

3. **Elliptic Envelope:**
The performance of this method on Approach 2 variables is found to be least effective. This would be due to the Non-Gaussian distribution nature of the data. Nevertheless, the model was trained with *'contamination'* of 0.1 and it predicted 1,51,996 instances as normal out of 172801. The same is depicted as Fig. 15.

```
# Let's predict how many data points has been identified as normal in attack data.
print(np.count_nonzero(Elliptic_Envelope.predict(P2_attack_PC)==1),'data points has been identified as normal')

151996 data points has been identified as normal
```

Fig. 15. Approach 2 Elliptic Envelope Normal

4 Results Analysis: Benchmarking and Discussion

After modelling of all three methods on both processes, their performance evaluation is carried out with the help of confusion matrix. ROC curve evaluation will be less effective in this case because the test data is imbalanced. A confusion matrix is a summary of prediction results on a classification problem [16].

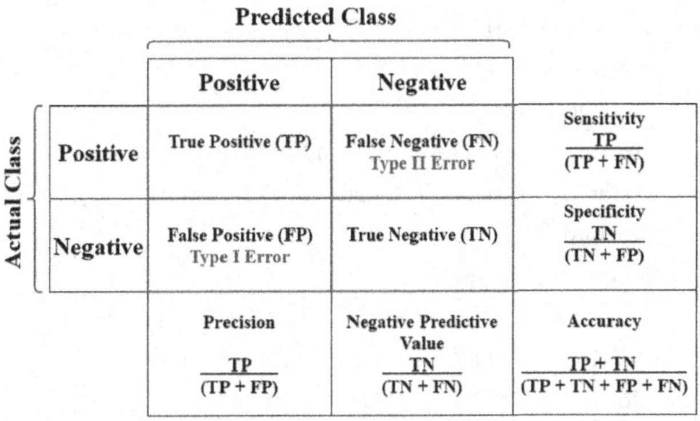

		Predicted Class		
		Positive	**Negative**	
Actual Class	**Positive**	**True Positive (TP)**	**False Negative (FN)** Type II Error	**Sensitivity** $\frac{TP}{(TP + FN)}$
	Negative	**False Positive (FP)** Type I Error	**True Negative (TN)**	**Specificity** $\frac{TN}{(TN + FP)}$
		Precision $\frac{TP}{(TP + FP)}$	**Negative Predictive Value** $\frac{TN}{(TN + FN)}$	**Accuracy** $\frac{TP + TN}{(TP + TN + FP + FN)}$

Fig. 16. Evaluation Metrics

A confusion matrix is a summary of prediction results on a classification problem (depicted in Fig. 16). The summarized analysis and benchmarking of both the models for corresponding processes is shown as Table 2 and 3.

4.1 Model Analysis

This section provides the model analysis for One class SVM, Isolation Forest and Elliptic Envelope computational processes.

Table 2. Model's Summary for Approach 1

Benchmarking of Trained Models: APPROACH 1				
One Class SVM				
Instance	Precision	Recall	F-1 score	Accuracy
Attack (−1)	0.25	0.48	0.33	0.88
Normal (1)	0.97	0.91	0.94	
Isolation Forest				
Instance	Precision	Recall	F-1 score	Accuracy
Attack (−1)	0.39	0.33	0.36	0.93
Normal (1)	0.96	0.97	0.96	
Elliptic Envelope				
Instance	Precision	Recall	F-1 score	Accuracy
Attack (−1)	0.21	0.32	0.26	0.89
Normal (1)	0.96	0.93	0.94	

Table 3. Model's Summary for Approach 2

Benchmarking of Trained Models: APPROACH 2				
One Class SVM				
Instance	Precision	Recall	F-1 score	Accuracy
Attack (−1)	0.32	0.48	0.38	0.91
Normal (1)	0.97	0.94	0.95	
Isolation Forest				
Instance	Precision	Recall	F-1 score	Accuracy
Attack (−1)	0.34	0.26	0.30	0.92
Normal (1)	0.96	0.97	0.96	
Elliptic Envelope				
Instance	Precision	Recall	F-1 score	Accuracy
Attack (−1)	0.07	0.13	0.09	0.84
Normal (1)	0.94	0.89	0.91	

1. **One Class SVM**
   - The second approach outperforms the first approach, especially in terms of the accuracy parameter.
   - The level of accuracy of both methods is higher in the case of normal incidents than in the case of attacks, indicating a tendency for conservative marking of anomalies.
   - Approach 2's One Class SVM performs better than Approach 1's in terms of accuracy, precision, and F1-score for both normal and attack cases.

2. **Isolation Forest**
   - Approach 2's Isolation Forest is way better than Approach 1's in all metrics for both normal and attack cases.
   - Among the three methods, Isolation Forest has shown the highest precision, recall, F1-score, and accuracy values in comparison with One Class SVM and Elliptic Envelope.
3. **Elliptic Envelope**
   - Eventually, the elliptic envelope has a lower performance than the one-class SVM and isolation forest for both approaches.
   - It represents the weakest performance of precision, recall, and F1-score in attack detection.

4.2 Consolidated Performance Analysis

Elliptic Envelope isn't effective for any process and even poor for Approach 2. It is found that deciding the better method between OC-SVM and Isolation Forest for Approach 1 is grinding. Where Isolation Forest gave high precision, on the other hand, OC-SVM have shown high recall. The decision in such a situation is user, as well as, problem dependent. Based on the desired constraints, priority will be given to an algorithm. Somewhat similar ambiguity also found in Approach 2. However, OC-SVM had outrun Isolation Forest with some differences in this case.

In summary, Isolation Forest performs consistently well across both approaches, demonstrating its effectiveness in anomaly detection. Approach 2 generally outperforms Approach 1, highlighting the importance of the chosen features or pre-processing steps in the anomaly detection process.

5 Conclusion and Future Work

This research work demonstrates that Machine Learning models can serve as an effective additional security layer for detecting process abnormalities. However, several critical factors must be addressed before and during the training of these models. The sensitivity of these algorithms to input data (training data) necessitates careful definition of normal operational parameters. The selection of variable ranges in the training data must be aligned with this definition. Furthermore, the internal parameters of these methods significantly influence the predictive outcomes, requiring iterative experimentation to determine the optimal parameter combinations. For instance, the Elliptic Envelope technique is recommended for training on low-dimensional data that may conform to a Gaussian distribution. The One Class SVM algorithm, known for its high sensitivity to outliers, is best applied to datasets from which outliers have been removed, making it particularly suitable for novelty detection tasks.

Future work is expected to explore the application of these models across various process industries by training on standardized operational data within defined system boundaries. There is potential for these models to be deployed in live monitoring systems, where they would process sensor data, apply transformation functions, and classify process states as normal or anomalous. Additionally, the Local Outlier Factor method, available in Scikit-Learn, warrants further investigation for its applicability in this context.

References

1. Tyagi, A.K., Sreenath, N.: Cyber physical systems: analyses, challenges and possible solutions. Internet Things Cyber-Phys. Syst. **1**, 22-33 (2021). https://doi.org/10.1016/jiotcps.2021.12.002, https://www.sciencedirect.com/science/article/pii/S2667345221000055. ISSN 2667-3452
2. Chandani, P., Rajagopal, S., Bishnoi, A.K., Verma, V.: Cyber-physical system and AI strategies for detecting cyber attacks in healthcare. Int. J. Intell. Syst. Appl. Eng. **11**(8s), 55–61 (2023). https://ijisae.org/index.php/IJISAE/article/view/3021
3. Duo, W., Zhou, M., Abusorrah, A.: A survey of cyber attacks on cyber physical systems: recent advances and challenges. IEEE/CAA J. Autom. Sinica **9**(5), 784–800 (2022). https://doi.org/10.1109/JAS.2022.105548
4. Binnar, P., Bhirud, S., Kazi, F.: Security analysis of cyber physical system using digital forensic incident response. Cyber Secur. Appl. **2**, 100 034 (2024). https://doi.org/10.1016/jcsa.2023.100034, https://www.sciencedirect.com/science/article/pii/S2772918423000218. ISSN 2772-9184
5. Chou, J.-S., Telaga, A.S.: Real-time detection of anomalous power consumption. Renew. Sustain. Energy Rev. **33**, 400–411 (2014)
6. Zhang, J.E., Wu, D., Boulet, B.: Time series anomaly detection for smart grids: a survey. In: 2021 IEEE Electrical Power and Energy Conference (EPEC), pp. 125–130. IEEE (2021)
7. Pang, G., Shen, C., Cao, L., Hengel, A.V.D.: Deep learning for anomaly detection: a review. ACM Comput. Surv. (CSUR) **54**(2), 1–38 (2021)
8. Gharibian, F., Ghorbani, A.A.: Comparative study of supervised machine learning techniques for intrusion detection. In: Fifth Annual Conference on Communication Networks and Services Research (CNSR 2007), pp. 350–358 (2007). https://doi.org/10.1109/CNSR.2007.22.
9. Fleizach, C., Fukushima, S.: A naive bayes classifier on 1998 KDD cup. Department of Computer Science Engineering, University of California, Los Angeles, CA, USA, Technical report (1998)
10. Taormina, R., Galelli, S.: Deep-learning approach to the detection and localization of cyber-physical attacks on water distribution systems. J. Water Resour. Plann. Manag. **144**(10), 04 018 065 (2018)
11. Wang, H., Zhang, G., Mingjie, E., Sun, N.: A novel intrusion detection method based on improved SVM by combining PCA and PSO. Wuhan Univ. J. Nat. Sci. **16**, 409–413 (2011)
12. Feng, C., Palleti, V.R., Mathur, A., Chana, D.: A systematic framework to generate invariants for anomaly detection in industrial control systems. In: NDSS, pp. 1–15 (2019)
13. Karnouskos, S.: Stuxnet worm impact on industrial cyber-physical system security. In: IECON 2011-37th Annual Conference of the IEEE Industrial Electronics Society, pp. 4490–4494. IEEE (2011)

14. Lin, Q., Adepu, S., Verwer, S., Mathur, A.: Tabor: a graphical model-based approach for anomaly detection in industrial control systems. In: Proceedings of the 2018 on Asia Conference on Computer and Communications Security, pp. 525–536 (2018)
15. Mathur, A.P., Tippenhauer, N.O.: Swat: a water treatment testbed for research and training on ICS security. In: 2016 International Workshop on Cyberphysical Systems for Smart Water Networks (CySWater), pp. 31–36. IEEE (2016)
16. Saito, T., Rehmsmeier, M.: The precision-recall plot is more informative than the ROC plot when evaluating binary classifiers on imbalanced datasets. PLoS ONE **10**(3), e0118432 (2015)

Author Index

S. Tiwari et al. (Eds.): AI4S 2024, CCIS 2243, pp. 247–248, 2025.
https://doi.org/10.1007/978-3-031-81369-6

Printed in the United States
by Baker & Taylor Publisher Services